AI and Us

Unraveling the Threads of Human
Connection

M.A. Gorre

Contents

Introduction   V

1. Human-AI Interaction   1

2. The Rise of Artificial Intelligence   7

3. Historical Evolution of AI   13

4. Types of AI   17

5. AI in Modern Society   22

6. Siri: Apple's Voice Assistant   28

7. Alexa: Amazon's Virtual Assistant   35

8. Google Assistant: Google's AI Companion   41

9. Cortana: Microsoft's Virtual Assistant   47

10. Bixby: Samsung's AI Voice Assistant   52

11. Other Virtual Assistants   57

12. How Virtual Assistants Work: The Power of Natural   64
    Language Processing (NLP)

13. How Speech Recognition Works   69

14. Machine Learning and Training   73

15. Cloud Computing and Data Processing   77

16. User Experience Design (UX Design)   85

17. Ethical considerations   90

18. Privacy Concerns   94

19. Virtual Assistants: Transforming Lives in Smart Homes 99

20. The Emotional Connection: 105

21. Challenges and Controversies Surrounding AI   110

22. The Future of AI and Human Relationships   115

23. References   120

24. Conclusion   123

Introduction

In the not-so-distant past, the idea of talking to a computer and expecting a meaningful response was the stuff of science fiction. The concept of Artificial Intelligence (AI) was confined to the realms of imagination, a futuristic notion that seemed eons away from our reality. But as we stand at the precipice of the third decade of the 21st century, AI has not only arrived but has become an integral part of our daily lives, altering how we interact with technology and, in many ways, with each other.

This book, "AI and Us: Unraveling the Threads of Human Connection," aims to delve deep into the transformative power of AI-driven virtual assistants and their profound influence on how we connect with technology, with one another, and even with ourselves. It's a journey into the heart of the digital age, where algorithms and neural networks are not just lines of code but companions, confidantes, and sometimes even catalysts for introspection.

In "AI and Us: Unraveling the Threads of Human Connection," we embark on a profound exploration of the impact of artificial intelligence (AI) on our lives, relationships, and the way we interact with technology. This book delves into the intricate web of AI and its role as a companion, assistant, and even confidant in our daily existence.

Virtual Companions Unveiled:

Discover the world of virtual assistants, including Siri, Alexa, Google Assistant, Cortana, and Bixby. Learn about the unique functions they serve and what sets them apart in our tech-savvy society.

Beneath the Surface:

Peel back the curtain to understand the inner workings of virtual assistants. From Natural Language Processing (NLP) to speech recognition, explore the technology that powers these AI marvels.

Human-AI Interaction Explored:

Delve into the nuances of human-AI interactions. From conversational interfaces to user experience design, we navigate the ethical considerations and privacy concerns that arise in our AI-driven world.

Daily Life Transformed:

Uncover how virtual assistants have seamlessly integrated into various aspects of daily life. From controlling smart homes to enhancing professional tasks, from healthcare applications to aid in education, explore the multifaceted impact of AI companions.

Emotional Bonds with AI:

Explore the captivating concept of emotional connections with AI. Delve into the human tendency to form attachments, the emotional responses evoked by AI interactions, and the intriguing idea of AI as companions.

Challenges and Controversies:

Confront the challenges and controversies in the world of AI. Engage in candid discussions about dependence on AI, job displacement, data privacy, security concerns, and bias in AI algorithms.

The Future Unveiled:

Gaze into the crystal ball of technology as we examine potential AI advancements, outline future scenarios of AI-human interactions, and discuss the necessity of ethical guidelines and regulations.

Personal Insights and Experiences:

To add a personal touch, the book features personal experiences with AI, interviews with experts, and stories from readers. It's a platform for sharing the human side of our AI-driven world.

"AI and Us: Unraveling the Threads of Human Connection" is a comprehensive exploration that navigates the evolving landscape of AI and its profound impact on human relationships and society. Join us on this thought-provoking journey through the world of artificial intelligence.

Conclusion:

In the concluding chapter, we'll summarize the key takeaways from our journey through AI and human relationships. We'll reflect on the profound changes and look ahead to the ever-evolving landscape of technology and its impact on our lives.

The adventure begins here, in the pages of "AI and Us: Unraveling the Threads of Human Connection," where we'll embark on a voyage through the realms of AI and human connection, navigating the seas of technology and exploring the uncharted territories of our digital future. Join me in exploring the profound, curious, and transformational world of AI and human relationships.

Human-AI Interaction

H uman-AI interaction and conversational interfaces represent a transformative aspect of technology, enabling natural and intuitive communication between humans and artificial intelligence systems. In this exploration, we'll delve into the significance of human-AI interaction, the role of conversational interfaces, and their real-world applications.

Human-AI Interaction: Bridging the Gap

Human-AI interaction is a multidisciplinary field that focuses on designing interfaces and interactions between humans and AI systems. It aims to make these interactions as seamless, effective, and user-friendly as possible. The ultimate goal is to bridge the gap between human cognitive capabilities and AI's computational power, enabling humans to harness the potential of AI for various tasks.

Key Aspects of Human-AI Interaction:

1. **Natural Language Processing (NLP):** NLP is at the core of human-AI interaction, allowing AI systems to understand and generate human language. This capability is essential for creating conversational interfaces.

2. **User-Centered Design:** The design of AI interfaces must prioritize the user's needs, preferences, and limitations. This involves creating interfaces that are intuitive and easy to use.

3. **Contextual Understanding:** Effective human-AI interaction requires AI systems to understand the context of a conversation, considering previous interactions and user intents.

4. **User Feedback:** Continuous feedback loops enable AI systems to learn from user interactions, improving their performance and understanding over time.

Conversational Interfaces: A Gateway to Human-AI Interaction

Conversational interfaces are a prominent application of human-AI interaction. They are AI-powered systems designed to facilitate natural and conversational interactions between humans and computers. These interfaces leverage NLP, machine learning, and other AI technologies to understand, interpret, and respond to spoken or written language.

Key Components of Conversational Interfaces:

1. **Speech Recognition:** Conversational interfaces employ speech recognition technology to transcribe spoken language into text, enabling voice-based interactions.

2. **Text-to-Speech (TTS):** TTS technology converts text responses into natural-sounding speech, making interactions more human-like.

3. **Natural Language Understanding (NLU):** NLU algorithms extract meaning from user inputs, identifying intents, entities, and context.

4. **Dialogue Management:** Dialogue management systems orchestrate conversations, ensuring coherent and context-aware interactions.

Applications of Conversational Interfaces:

1. **Virtual Assistants:** Virtual assistants like Siri, Alexa, Google Assistant, and Cortana are perhaps the most recognizable conversational interfaces. They assist users with tasks, answer questions, and perform actions based on voice commands.

2. **Chatbots:** Chatbots are text-based conversational interfaces often used in customer support, e-commerce, and various online services. They engage users in text-based conversations to provide information or assistance.

3. **Voice-Activated Devices:** Devices like smart speakers and IoT appliances use conversational interfaces to allow users to control functions, get information, or perform tasks through voice commands.

4. **Healthcare:** Conversational interfaces are increasingly used in healthcare for tasks like appointment scheduling, medication reminders, and symptom assessment.

5. **Education:** AI-driven chatbots and virtual tutors assist students with questions, explanations, and personalized learning experiences.

6. **Business and Productivity:** Chatbots and virtual assistants help businesses streamline operations, manage appointments, and provide customer support.

The Significance of Conversational Interfaces:

1. **Enhanced User Experience:** Conversational interfaces offer a more natural and user-friendly way to interact with technology, reducing the learning curve and making AI accessible to a broader audience.

2. **Efficiency and Productivity:** These interfaces can streamline tasks and provide information rapidly, enhancing efficiency and productivity in various domains.

3. **Accessibility:** Conversational interfaces are inclusive, as they allow people with disabilities or language barriers to interact with technology more effectively.

4. **Data-Driven Insights:** Conversational interfaces generate valuable data about user preferences and behaviors, enabling organizations to improve their products and services.

5. **Automation:** These interfaces can automate routine tasks, freeing up human resources for more complex and creative endeavors.

Challenges in Conversational Interfaces:

1. **Understanding Ambiguity:** Human language is often ambiguous, requiring AI systems to disambiguate and interpret user intent accurately.

2. **Privacy Concerns:** Conversational interfaces handle sensitive information, raising privacy and security concerns that must be addressed.

3. **Bias and Fairness:** AI models used in conversational interfaces can inherit biases from training data, leading to unfair

or discriminatory responses.

4. **Integration:** Integrating conversational interfaces with existing systems and workflows can be complex and require substantial development effort.

The Future of Human-AI Interaction and Conversational Interfaces:

The field of human-AI interaction and conversational interfaces continues to evolve rapidly. Future developments may include:

1. **Multimodal Interaction:** Combining voice, text, and visual inputs for more immersive and versatile interactions.

2. **Emotion Recognition:** AI systems that can detect and respond to user emotions, enhancing the emotional intelligence of conversational interfaces.

3. **Cross-Platform Integration:** Seamless interaction across various devices and platforms, making AI accessible in diverse contexts.

4. **Ethical AI:** Efforts to address biases, transparency, and accountability in conversational interfaces.

In conclusion, conversational interfaces are at the forefront of human-AI interaction, revolutionizing the way we engage with technology. As AI technologies advance and user expectations evolve, these interfaces will continue to play a pivotal role in simplifying interactions, enhancing user experiences, and driving innovation across industries. The future promises even more natural, intuitive, and context-aware conversations between humans and AI.

The Rise of Artificial Intelligence

T he rise of Artificial Intelligence (AI) is a story of human ingenuity and relentless pursuit of progress. It's a tale that spans
decades, with humble beginnings in the realm of mathematics and
the lofty aspirations of replicating human intelligence in machines. To
truly understand the significance of AI today, we must first journey
back in time to its inception and trace its evolution through the annals
of scientific discovery.

From Turing to Neural Networks: The Early Days

The seeds of AI were sown in the fertile grounds of mathematics
and logic. Alan Turing, the brilliant British mathematician, laid the
foundation for AI with his pioneering work on computability and
the concept of a "universal machine" in the 1930s. Turing's ideas were
pivotal in shaping the notion of a machine that could mimic human
thought processes.

The term "Artificial Intelligence" itself was coined by John Mc-
Carthy in 1956, marking the birth of a new field of research. Mc-

Carthy, along with a group of visionary researchers, organized the Dartmouth Workshop, a seminal event that convened experts to explore the possibilities of machine intelligence. This workshop is considered the birthplace of AI as a formal discipline.

However, the early days of AI were fraught with optimism and naivety. The pioneers of AI believed that they could quickly replicate human-level intelligence in machines. These early expectations often exceeded the technological capabilities of the time, leading to what would later be known as the "AI winter" – periods of reduced funding and disillusionment with the field.

Symbolic AI and Rule-Based Systems

The initial approach to AI, known as "symbolic AI," relied heavily on rule-based systems and symbolic representation of knowledge. Researchers attempted to encode human expertise into computer programs using explicit rules and logical reasoning. This approach found success in narrow domains such as chess, where IBM's Deep Thought became the first computer to defeat a reigning world champion.

However, symbolic AI encountered limitations when applied to complex, real-world problems. The world is often messy and unpredictable, and rule-based systems struggled to handle the ambiguity and uncertainty inherent in many tasks.

The AI Winter and the Emergence of Machine Learning

As the limitations of symbolic AI became apparent, the field entered a period of stagnation known as the AI winter. Funding dried up, and AI research struggled to gain traction. But beneath the surface, a revolution was brewing.

Machine learning, a subfield of AI that focused on developing algorithms that could learn from data, began to gain prominence. Researchers like Arthur Samuel, who coined the term "machine learning" in 1959, paved the way for a new paradigm in AI. Instead of hand-crafting rules, machine learning algorithms could autonomously discover patterns and make predictions from data.

The Renaissance of Deep Learning

The turning point for AI came with the resurgence of neural networks and deep learning. Neural networks, inspired by the structure of the human brain, had been around since the 1940s but fell out of favor during the AI winter due to computational limitations. However, advances in hardware and the availability of vast datasets rekindled interest in neural networks.

In 2012, the deep learning community was thrust into the spotlight when a deep convolutional neural network named AlexNet won the ImageNet competition, significantly surpassing human performance in image classification. This breakthrough demonstrated the power of deep learning and its potential applications beyond image recognition.

Understanding AI

Now that we've explored the historical journey of AI, let's turn our attention to understanding what AI is and how it operates in the modern world.

Defining Artificial Intelligence

Artificial Intelligence refers to the development of computer systems that can perform tasks typically requiring human intelligence. These tasks encompass a wide range of activities, including but not limited to:

1. **Machine Learning:** AI systems can learn from data and improve their performance over time. They use algorithms to recognize patterns, make predictions, and make decisions based on input.

2. **Natural Language Processing (NLP):** NLP enables AI to understand, interpret, and generate human language. It

powers virtual assistants like Siri and chatbots that can engage in human-like conversations.

3. **Computer Vision:** AI can analyze and interpret visual information from images and videos. This capability is crucial for applications like facial recognition and autonomous vehicles.

4. **Robotics:** AI-driven robots can perform tasks in the physical world, from manufacturing to healthcare assistance.

5. **Expert Systems:** These are AI systems that emulate the decision-making abilities of a human expert in a specific domain, such as medical diagnosis or financial analysis.

Machine Learning: The Heart of AI

At the core of many AI applications is machine learning. Machine learning is a subset of AI that focuses on the development of algorithms capable of learning from and making predictions or decisions based on data. This process involves the following key components:

1. **Data:** Machine learning algorithms require large amounts of data to train. This data can be structured (e.g., databases) or unstructured (e.g., text, images, audio).

2. **Training:** During the training phase, the algorithm learns patterns and relationships in the data. It adjusts its internal parameters to optimize its performance on specific tasks.

3. **Inference:** Once trained, the machine learning model can make predictions or decisions when presented with new, unseen data. This is known as inference.

4. **Feedback Loop:** Machine learning systems often have a

feedback loop where they continuously learn from new data and improve their performance over time.

Machine learning encompasses various techniques, including supervised learning (where the algorithm learns from labeled examples), unsupervised learning (where it discovers patterns without labeled data), and reinforcement learning (where an agent learns to make decisions through trial and error).

Natural Language Processing (NLP) and Computer Vision

Two prominent AI subfields are Natural Language Processing (NLP) and Computer Vision.

NLP focuses on enabling computers to understand, interpret, and generate human language. This technology is the backbone of virtual assistants like Siri and chatbots that engage in human-like conversations. NLP also powers language translation, sentiment analysis, and text summarization.

Computer Vision, on the other hand, enables computers to interpret visual information from the world, including images and videos. It's vital for facial recognition, object detection, autonomous vehicles, and medical image analysis.

The Ethical and Societal Implications of AI

Understanding AI also involves acknowledging the ethical and societal implications of this technology. As AI becomes more integrated into our lives, questions of fairness, transparency, privacy, and bias arise. Ensuring that AI systems are developed and deployed responsibly is a critical concern for governments, organizations, and researchers.

Conclusion

The rise of Artificial Intelligence is a testament to human curiosity, innovation, and determination. From its humble beginnings as a mathematical concept to its current state as a transformative technology, AI has come a long way. It has evolved from symbolic AI to the renaissance of deep learning, shaping the digital landscape as we know it.

Understanding AI is about recognizing its capabilities, its reliance on data and learning, and its role in revolutionizing various industries. It's also about acknowledging the ethical and societal responsibilities that come with this powerful technology. As we embark on this exploration of AI and its impact on human relationships, we must keep in mind that AI is not just a tool; it's a force that's reshaping the way we live, work, and connect with the world around us.

Historical Evolution of AI

The Birth of AI: The Dream Begins (1940s-1950s)

The story of AI begins in the 1940s and 1950s, a time when scientists and mathematicians started to dream of creating machines that could mimic human intelligence. One of the earliest pioneers was Alan Turing, a brilliant British mathematician. In the 1930s, Turing developed the concept of a "universal machine" that could simulate any other machine, laying the theoretical groundwork for AI.

In 1956, John McCarthy, along with a group of researchers, organized the Dartmouth Workshop, where they coined the term "Artificial Intelligence" and set out to explore the possibilities of creating intelligent machines. This event marked the birth of AI as a formal field of study.

The Early Years: Symbolic AI and Rule-Based Systems (1950s-1960s)

In the early years of AI research, the predominant approach was "symbolic AI." Researchers attempted to replicate human intelligence by encoding knowledge and logical rules into computer programs. These rule-based systems were good at tasks that required logical reasoning but struggled with uncertainty and complex real-world problems.

During this period, early AI systems could solve mathematical problems and play chess at a basic level. For example, the IBM computer "Deep Thought" became the first to defeat a reigning world chess champion in 1989. However, the dream of creating general-purpose intelligent machines remained elusive.

The AI Winter: Setbacks and Realizations (1970s-1980s)

The AI field faced significant challenges and setbacks during the 1970s and 1980s, often referred to as the "AI winter." Unrealistic expectations led to disappointment, and funding for AI research dwindled. Many realized that symbolic AI, based on explicit rules, couldn't easily handle the complexity and ambiguity of the real world.

A New Paradigm Emerges: Machine Learning (1950s-1980s)

Amid the challenges, a new paradigm began to gain traction: machine learning. Machine learning focused on creating algorithms that could learn from data rather than relying solely on pre-defined rules. Arthur Samuel, a pioneer in the field, coined the term "machine learning" in 1959.

Machine learning algorithms could autonomously discover patterns and make predictions from data. While these early machine learning methods were limited by computing power and data availability, they laid the foundation for later breakthroughs.

The Renaissance of Neural Networks (1980s-1990s)

One of the most significant developments during this period was the resurgence of neural networks. Neural networks, inspired by the structure of the human brain, had been around since the 1940s but were largely overshadowed by symbolic AI. However, advances in computing power and the availability of large datasets breathed new life into neural networks.

In 2012, a deep convolutional neural network named "AlexNet" won the ImageNet competition, significantly surpassing human performance in image classification. This marked a turning point, showcasing the potential of deep learning, a subset of machine learning that uses deep neural networks.

The AI Revolution: Deep Learning and Breakthroughs (2010s-Present)

The past decade has witnessed an AI revolution, driven by deep learning and neural networks. Deep learning models, including convolutional neural networks (CNNs) and recurrent neural networks (RNNs), have achieved remarkable success in various domains.

Applications of Modern AI

- **Natural Language Processing (NLP):** NLP models like BERT and GPT-3 can understand and generate human language, enabling chatbots, language translation, and content generation.

- **Computer Vision:** AI systems can now recognize objects, faces, and even emotions in images and videos, enabling applications like facial recognition, self-driving cars, and medical image analysis.

- **Reinforcement Learning:** AI agents, trained through reinforcement learning, have excelled in complex tasks, such as playing video games and mastering board games like Go.

- **Healthcare:** AI is revolutionizing healthcare with applications in medical image analysis, disease diagnosis, and drug discovery.

The Future of AI: Ethical and Societal Considerations

As AI continues to advance, it raises important ethical and societal questions. Issues like bias in AI algorithms, data privacy, and the impact on employment must be carefully addressed. Ensuring that AI is developed and deployed responsibly is crucial as we move forward.

In conclusion, the historical evolution of AI is a journey from humble beginnings to groundbreaking achievements. From rule-based systems to machine learning and deep learning, AI has come a long way. Today, AI touches nearly every aspect of our lives, and its potential for the future is limitless. As we navigate this AI-driven world, understanding its history helps us appreciate the challenges and opportunities that lie ahead. AI is no longer just a dream; it's a powerful reality shaping the 21st century.

Types of AI

From Narrow to Superintelligent

Artificial Intelligence (AI) is a diverse field, with different levels of AI capabilities. These range from narrow, task-specific AI to the hypothetical superintelligent AI that surpasses human intelligence. Let's explore the key types of AI, their characteristics, and real-world applications.

1. Narrow or Weak AI (ANI - Artificial Narrow Intelligence)

Narrow AI, also known as Weak AI, is AI designed and trained for a specific task or a set of closely related tasks. It operates under a narrow domain and lacks general intelligence. While it can perform tasks at or above human level within its specific domain, it doesn't possess the ability to transfer knowledge or skills to unrelated tasks. Some characteristics of Narrow AI include:

- **Task-Specific:** Narrow AI is designed for a single purpose or a limited set of tasks. For example, virtual assistants like Siri and Alexa are specialized in natural language understanding and providing information.

- **Data-Dependent:** It relies heavily on data to make decisions or perform tasks. Machine learning and deep learning techniques are commonly used in narrow AI to analyze data and make predictions.

- **Limited Adaptability:** It cannot adapt to tasks outside its predefined domain. For instance, a narrow AI designed for image recognition cannot suddenly start translating languages.

- **Examples:** Chatbots, recommendation systems, image and speech recognition, autonomous vehicles (to some extent), and virtual personal assistants.

2. General or Strong AI (AGI - Artificial General Intelligence)

General AI, or Strong AI, represents a higher level of intelligence. It possesses human-like cognitive abilities and can understand, learn, and apply knowledge across a broad range of tasks. AGI doesn't exist yet, and creating it is one of the long-term goals of AI research. Some characteristics of General AI include:

- **General Problem Solving:** AGI can tackle a wide variety of problems and tasks, similar to how a human would approach them.

- **Learning and Adaptation:** It has the ability to learn from experience, generalize knowledge, and apply it to new situations.

- **Common Sense Reasoning:** AGI can exhibit common sense reasoning and exhibit a deep understanding of context, making it versatile in various domains.

- **Self-awareness:** A true AGI might possess self-awareness and consciousness, although this remains speculative.

3. Artificial Narrow Superintelligence (ANSI)

Artificial Narrow Superintelligence refers to AI systems that surpass human intelligence in specific, narrow domains. While they excel in these areas, they lack the broad, general intelligence of AGI. ANSI can perform tasks that would be impossible for humans to accomplish within their respective domains. Characteristics of ANSI include:

- **Domain-Specific Excellence:** It is designed to outperform humans in narrowly defined domains, such as chess, Go, or complex simulations.

- **Rapid Learning:** ANSI can rapidly acquire and apply new knowledge within its specialized domain, often surpassing human capabilities.

- **Lack of Generalization:** Despite its exceptional performance in specific areas, ANSI doesn't possess the ability to transfer its expertise to unrelated domains.

- **Examples:** Deep Blue, the chess-playing computer, is an example of ANSI that excelled in the domain of chess but lacked general intelligence.

4. Artificial General Superintelligence (AGSI)

Artificial General Superintelligence represents the hypothetical pinnacle of AI development. AGSI would surpass not only human intelligence but also encompass superhuman intelligence across all possible domains. It would be capable of understanding, learning, and innovating in any field. Characteristics of AGSI include:

- **Unlimited Learning and Adaptation:** AGSI could learn and adapt to any domain or task, no matter how complex or diverse.

- **Creative Problem Solving:** It would exhibit creativity, innovation, and the ability to solve novel problems beyond human comprehension.

- **Autonomy and Self-Improvement:** AGSI could improve its own capabilities, leading to rapid self-enhancement and exponential growth in intelligence.

- **Ethical Considerations:** Achieving AGSI poses significant ethical and existential risks, as it could potentially surpass human control and understanding.

Real-World Applications and Implications

In the real world, the majority of AI systems fall under the category of Narrow AI. These systems are employed in various applications, from virtual assistants that aid in natural language processing to recommendation systems that personalize content. Narrow AI is also used in industries like healthcare for medical diagnosis, in finance for fraud detection, and in transportation for autonomous vehicles.

The development of AGI and AGSI remains a topic of ongoing research and speculation. Achieving AGI, if it's possible, would be a monumental milestone in AI history, with profound implications for society, ethics, and technology. Concerns about the ethical and societal impact of superintelligent AI have led to discussions about control mechanisms, regulations, and responsible AI development.

In conclusion, the field of Artificial Intelligence encompasses various types, from Narrow AI designed for specific tasks to the theoretical realm of superintelligent AI. While Narrow AI is prevalent in our daily lives, the pursuit of AGI and AGSI represents the quest for achieving human-like and superhuman intelligence, respectively. As AI technology advances, the balance between innovation and ethical considerations becomes increasingly crucial for shaping our AI-driven future.

AI in Modern Society

Shaping the Future

Artificial Intelligence (AI) has emerged as a transformative force in modern society, revolutionizing how we live, work, and interact. This technology, once confined to science fiction, is now deeply integrated into our daily lives, influencing everything from healthcare to entertainment. In this exploration, we'll delve into the multifaceted role of AI in modern society, its applications, and the ethical challenges it raises.

AI in Healthcare

AI's influence in healthcare is profound, with applications spanning from diagnostics to personalized medicine:

1. **Medical Imaging:** AI-powered algorithms analyze medical images like X-rays and MRIs, aiding radiologists in detecting diseases such as cancer and identifying anomalies with high precision.

2. **Drug Discovery:** AI accelerates drug development by predicting potential drug candidates, simulating molecular interactions, and sifting through vast datasets for potential

treatments.

3. **Clinical Decision Support:** AI systems assist healthcare professionals in making diagnoses, suggesting treatment options, and predicting patient outcomes.

4. **Personalized Medicine:** AI leverages genetic and patient data to tailor treatments to individuals, optimizing therapy effectiveness and minimizing side effects.

5. **Healthcare Management:** AI improves administrative tasks, including appointment scheduling, billing, and resource allocation.

AI in Education

AI is reshaping the education landscape, enhancing learning experiences and personalizing education:

1. **Adaptive Learning:** AI-driven platforms adapt to individual student needs, offering tailored lessons and assessments.

2. **Virtual Tutors:** AI-powered virtual tutors provide immediate feedback, helping students grasp complex concepts.

3. **Language Learning:** Language learning apps use AI to improve pronunciation and language understanding.

4. **Automated Grading:** AI automates grading, reducing the administrative burden on teachers and enabling timely feedback.

5. **Educational Content Creation:** AI generates educational content, from writing essays to designing interactive materials.

AI in Finance

The financial sector relies on AI for data analysis, risk assessment, and fraud prevention:

1. **Algorithmic Trading:** AI-driven algorithms execute high-frequency trades, capitalizing on market trends.

2. **Risk Assessment:** AI assesses borrower creditworthiness, insurance risks, and investment portfolios.

3. **Fraud Detection:** AI detects fraudulent transactions in real-time, safeguarding financial institutions and customers.

4. **Customer Service:** Chatbots and virtual assistants handle routine customer inquiries, improving service efficiency.

5. **Predictive Analytics:** AI predicts market trends, enabling more informed investment decisions.

AI in Transportation

AI plays a pivotal role in modern transportation, particularly in the development of autonomous vehicles:

1. **Self-Driving Cars:** AI algorithms power autonomous vehicles, offering the promise of safer and more efficient transportation.

2. **Traffic Management:** AI optimizes traffic flow, reducing congestion and enhancing urban mobility.

3. **Navigation:** GPS systems incorporate AI for real-time traffic updates and route optimization.

4. **Public Transport:** AI-driven apps provide information on public transit schedules and service disruptions.

5. **Delivery Services:** AI is used for route planning and package sorting in delivery logistics.

AI in Entertainment and Media

AI transforms the entertainment industry through content creation and personalization:

1. **Content Recommendation:** Streaming platforms use AI to recommend movies, music, and shows based on user preferences.

2. **Content Generation:** AI generates music, art, and even news articles, blurring the lines between human and machine creativity.

3. **Video Editing:** AI automates video editing tasks, reducing production time and costs.

4. **Gaming:** AI powers non-player characters (NPCs), providing dynamic and challenging gaming experiences.

5. **Virtual Reality (VR) and Augmented Reality (AR):** AI enhances immersive experiences in VR and AR applications.

AI in Retail

Retailers leverage AI to enhance customer experiences, streamline operations, and optimize inventory management:

1. **Personalized Shopping:** AI-driven recommendations and personalized shopping experiences improve customer satisfaction.

2. **Inventory Management:** AI forecasts demand, reducing stockouts and overstock situations.

3. **Supply Chain Optimization:** AI optimizes supply chain logistics, minimizing delays and costs.

4. **Chatbots:** Chatbots handle customer inquiries and assist in online shopping.

5. **Pricing Optimization:** AI adjusts prices dynamically based on demand and market conditions.

AI in Cybersecurity

AI bolsters cybersecurity efforts by detecting threats and enhancing network security:

1. **Threat Detection:** AI analyzes network traffic for anomalies and identifies potential security breaches.

2. **Malware Detection:** AI-powered antivirus programs detect and mitigate malware threats.

3. **Authentication:** AI enhances user authentication with biometrics and behavioral analysis.

4. **Data Protection:** AI encrypts sensitive data and monitors access to prevent unauthorized breaches.

5. **Incident Response:** AI automates incident response processes, minimizing damage in the event of a breach.

Ethical Considerations in AI

The widespread adoption of AI in modern society brings ethical concerns to the forefront:

1. **Bias and Fairness:** AI algorithms can inherit biases from training data, leading to unfair outcomes and discrimination.

2. **Privacy:** AI systems process vast amounts of personal data, raising concerns about data privacy and surveillance.

3. **Transparency:** The "black box" nature of some AI models makes it challenging to understand their decision-making processes.

4. **Accountability:** Determining responsibility for AI decisions, especially in critical applications like healthcare and autonomous vehicles, poses challenges.

5. **Job Displacement:** Automation through AI may lead to job displacement, requiring strategies for workforce transition.

The Future of AI in Modern Society

As AI continues to advance, its role in modern society will become increasingly prominent. The future holds possibilities for AI-powered healthcare diagnostics, autonomous transportation networks, and AI-augmented creativity in fields like art and music. However, managing the ethical implications and ensuring responsible AI development will be paramount as we navigate this transformative era.

In conclusion, AI has permeated modern society, offering opportunities for innovation and efficiency across various sectors. While its potential is vast, addressing ethical challenges and ensuring responsible AI adoption will be essential to harnessing the benefits of this transformative technology for the betterment of society.

Siri: Apple's Voice Assistant

Virtual Assistants and Their Functions: Unveiling the Power of Siri

In an increasingly digital world, virtual assistants have become an integral part of our daily lives. These AI-driven companions offer convenience, efficiency, and even a touch of personality. Siri, Apple's voice assistant, is one of the most well-known virtual assistants, providing users with a wide range of functions and capabilities. In this exploration, we'll unravel the world of virtual assistants, with a spotlight on Siri, examining its evolution, functions, and impact on our digital interactions.

The Rise of Virtual Assistants

Virtual assistants represent a significant milestone in the evolution of human-computer interaction. They are the result of advancements in Artificial Intelligence (AI), Natural Language Processing (NLP), and speech recognition technologies. Virtual assistants aim to understand and respond to human language, providing users with information, performing tasks, and even engaging in conversations.

The concept of virtual assistants had its roots in science fiction, where intelligent, voice-activated computer systems like "HAL 9000" from "2001: A Space Odyssey" captured the imagination of audiences. However, it wasn't until the advent of modern AI that virtual assistants became a reality.

Siri's Genesis: A Brief History

Siri, Apple's virtual assistant, had its beginnings as an independent project by SRI International, a California-based research institute. In 2007, Siri, Inc. was founded by Dag Kittlaus, Adam Cheyer, and Tom Gruber, with the aim of creating a voice-activated personal assistant. The name "Siri" is of Norwegian origin and means "beautiful woman who leads you to victory."

In April 2010, Apple acquired Siri, Inc., and Siri was integrated into the iOS ecosystem, making its debut with the release of the iPhone 4S in October 2011. This marked a significant moment in the world of technology, as Siri became one of the first widely adopted virtual assistants available to the masses.

Understanding Siri's Functions

Siri's functions encompass a wide range of tasks and capabilities, making it a versatile virtual assistant. Let's explore some of the key functions that Siri offers:

1. Voice Commands and Dictation:

At its core, Siri is designed for voice interaction. Users can activate Siri by saying, "Hey Siri" or pressing a designated button on their Apple devices. Siri can perform tasks based on voice commands, from setting alarms and sending texts to providing weather forecasts and playing music.

2. Information Retrieval:

Siri serves as a knowledge repository, capable of answering questions and providing information on a wide array of topics. Users can ask Siri about historical facts, sports scores, movie recommendations, and more. Siri's ability to fetch information from the web and present concise responses has made it a valuable source of knowledge.

3. Device Control:

Siri seamlessly integrates with Apple devices and can control various functions, such as adjusting screen brightness, toggling Wi-Fi settings, and activating "Do Not Disturb" mode. This functionality extends to smart home devices through Apple's HomeKit, allowing users to control lights, thermostats, and locks with voice commands.

4. Personal Assistant Tasks:

Siri excels at assisting with personal tasks. Users can set reminders, create calendar events, and manage to-do lists simply by asking Siri. This hands-free approach to task management streamlines users' daily routines.

5. Navigation and Directions:

Siri's integration with Apple Maps enables users to ask for directions, find nearby restaurants, and check traffic conditions. Siri provides turn-by-turn navigation, making it a valuable tool for travelers and commuters.

6. Messaging and Communication:

Siri simplifies communication by sending text messages, making phone calls, and even reading aloud incoming messages. This functionality enhances accessibility and safety while driving or multitasking.

7. Language Translation:

Siri's language capabilities extend beyond English, supporting numerous languages and dialects. Users can ask Siri to translate phrases or provide pronunciation assistance, making it a useful tool for language learners and travelers.

8. Entertainment and Media:

Siri enriches the entertainment experience by controlling music playback, identifying songs, and recommending movies and TV shows. It can even provide sports updates and scores for enthusiasts.

9. Accessibility Features:

Siri enhances accessibility for users with disabilities by performing tasks like turning on VoiceOver (a screen-reading feature), adjusting display settings, and providing voice-guided navigation.

10. Conversational AI:

Siri's conversational abilities have improved over the years. It can engage in more natural and context-aware dialogues, making interactions feel less transactional and more like conversations with a human assistant.

The Impact of Siri on Modern Society

Siri's impact on modern society is multifaceted, influencing how we interact with technology and integrate it into our daily routines:

1. Redefining Human-Computer Interaction:

Siri introduced a new era of voice-driven interaction with technology. The ability to speak naturally to a device and receive meaningful responses has changed how we search for information, manage tasks, and control our devices.

2. Accessibility and Inclusivity:

Siri's accessibility features have empowered individuals with disabilities, offering voice-driven assistance for tasks that may have been challenging otherwise. This inclusivity aligns with Apple's commitment to making technology accessible to all.

3. Integration into Ecosystems:

Siri's integration with Apple's ecosystem of devices and services has created a seamless and interconnected user experience. From iPhones to Macs and HomePods, Siri provides continuity across various Apple products.

4. Convenience and Efficiency:

Siri's ability to perform tasks with voice commands enhances convenience and efficiency. Whether it's setting reminders, sending texts, or controlling smart home devices, Siri simplifies everyday activities.

5. Personalization and Recommendations:

Siri's knowledge base and recommendation capabilities have made it a personalized assistant, providing tailored content suggestions and helping users discover new music, movies, and apps.

Ethical Considerations and Privacy

The use of virtual assistants like Siri also raises important ethical considerations, particularly regarding user privacy and data security. Apple has emphasized its commitment to user privacy by implementing features like on-device processing and anonymizing data. Siri interactions are designed to stay on the user's device whenever possible, minimizing the exposure of personal information.

However, the collection and processing of voice data for improving Siri's performance have sparked discussions about user consent and data transparency. Apple has introduced features that allow users to control their data sharing preferences and delete Siri recordings.

Challenges and Future Developments

While Siri has made significant strides in AI and voice recognition, there are ongoing challenges and opportunities for improvement:

> 1. **Accuracy and Context:** Siri's understanding of context and natural language remains an area for enhancement. Improving contextual awareness and reducing misinterpretations will make interactions even more seamless.

2. **Multilingual Support:** Expanding Siri's capabilities in languages and dialects worldwide will further its global accessibility and appeal.

3. **Third-Party Integration:** Siri continues to evolve as a platform for third-party app integration, allowing developers to create custom Siri actions and enhance its functionality.

4. **Privacy and Transparency:** Addressing user concerns regarding data privacy and maintaining transparency in data handling will remain crucial.

5. **AI Advancements:** Leveraging advancements in AI, machine learning, and NLP will enable Siri to better understand user intent and provide more accurate responses.

Conclusion

Siri, Apple's voice assistant, has transformed the way we interact with technology. Its functions encompass a wide range of tasks, from answering questions and managing tasks to controlling devices and providing entertainment. Siri's impact on modern society is evident in the convenience and accessibility it brings to users' lives.

As technology continues to advance, Siri's capabilities will evolve, offering more personalized and context-aware interactions. However, it's essential to navigate the ethical considerations surrounding virtual assistants, particularly in terms of data privacy and transparency.

Siri's journey from a research project to a household name exemplifies the power of AI to enhance human-computer interaction and redefine our digital experiences. In the ever-evolving landscape of virtual assistants, Siri remains a leading voice in the conversation, promising exciting developments in the years to come.

Alexa: Amazon's Virtual Assistant

Alexa: **Amazon's Virtual Assistant Transforming Our Lives**

In the age of digital transformation, virtual assistants have taken center stage, making everyday tasks more manageable and convenient. Among the pioneers of virtual assistant technology is Amazon's Alexa, a voice-activated AI assistant that has become a household name. In this exploration, we will delve into the world of Alexa, tracing its evolution, functions, and the profound impact it has had on our daily lives.

The Birth of Alexa

Alexa emerged from the innovation hub of Amazon, the e-commerce giant that had already disrupted the way we shop online. The journey of Alexa began with a team of Amazon engineers and developers in the early 2010s, driven by a vision to create a voice-activated, intelligent assistant that could simplify various tasks.

In November 2014, Amazon introduced the first Echo device, a cylindrical speaker equipped with Alexa. This marked the official launch of Alexa, which would later expand beyond Echo to other Amazon devices, third-party smart speakers, and even mobile apps.

Understanding Alexa's Functions

Alexa is renowned for its versatility, offering a wide array of functions and capabilities that make it a valuable addition to households, businesses, and various industries. Let's explore some of the key functions that Alexa provides:

1. Voice Commands and Natural Language Processing:

At the heart of Alexa's functionality is voice interaction. Users can activate Alexa by saying the wake word, "Alexa," followed by a command or question. Alexa leverages advanced Natural Language Processing (NLP) to understand and respond to spoken language, making interactions feel more intuitive.

2. Information Retrieval:

One of Alexa's primary functions is providing information and answering questions. Users can ask Alexa about the weather, news, historical facts, sports scores, and a wide range of general knowledge queries. Alexa fetches answers from the web and presents concise responses, effectively serving as a virtual encyclopedia.

3. Smart Home Control:

Alexa serves as a central hub for smart home devices, allowing users to control lights, thermostats, locks, cameras, and more. Its compatibility with various smart home ecosystems, including Amazon's own Echo ecosystem and third-party devices, makes it a powerful tool for home automation.

4. Entertainment and Media:

Alexa enhances entertainment experiences by controlling music playback, providing recommendations, and connecting to streaming services. Users can ask Alexa to play music, read audiobooks, and even provide updates on movies, TV shows, and sports events.

5. Productivity and Organization:

Alexa simplifies daily routines by setting alarms, timers, and reminders. It also integrates with calendar applications, helping users manage appointments and schedules. To-do lists and shopping lists can be created and managed through voice commands.

6. Navigation and Directions:

Alexa's integration with navigation services allows users to ask for directions, find nearby businesses, and check traffic conditions. It provides turn-by-turn directions for driving and walking.

7. Third-Party Skills and Customization:

One of Alexa's most distinctive features is its ability to expand its capabilities through third-party skills. Developers can create skills that add new functionalities, such as ordering food, playing interactive games, and providing fitness routines. Users can enable and customize skills based on their preferences.

8. Language Support:

Alexa is available in multiple languages and dialects, allowing users around the world to interact with it in their native tongue. Its multilingual capabilities make it accessible to a global audience.

9. Accessibility Features:

Alexa's accessibility features include voice control for individuals with mobility impairments, screen reading capabilities, and compatibility with Braille displays. These features make Alexa more inclusive and accessible to users with disabilities.

The Impact of Alexa on Modern Society

Alexa's introduction into modern society has been transformative, reshaping the way we interact with technology and integrate it into our daily lives:

1. Redefining Smart Homes:

Alexa has played a pivotal role in popularizing the concept of the smart home. Its ability to control various smart devices and provide centralized management has made home automation more accessible and user-friendly.

2. Voice-First Revolution:

Alexa is at the forefront of the voice-first revolution, where voice commands replace traditional touch and typing interactions. This shift has implications for user interfaces and the way we interact with technology.

3. Ecosystem Expansion:

The Echo ecosystem, driven by Alexa, has expanded to include a wide range of Echo devices, from smart speakers to displays and appliances. Alexa's integration with these devices creates a seamless and interconnected user experience.

4. Convenience and Efficiency:

Alexa's voice commands simplify tasks that would otherwise require manual input. Whether it's setting reminders, controlling lights, or playing music, Alexa enhances convenience and efficiency in users' daily routines.

5. Third-Party Innovation:

The open nature of the Alexa platform has spurred innovation from third-party developers who create skills and integrations, expanding Alexa's functionality and catering to diverse user needs.

6. Accessibility Advancements:

Alexa's accessibility features have made technology more inclusive, ensuring that individuals with disabilities can benefit from voice-driven interactions and smart home control.

Ethical Considerations and Privacy

The proliferation of virtual assistants like Alexa raises ethical concerns, particularly in terms of user privacy and data security. Amazon has taken steps to address these concerns by implementing features like voice profile recognition, which enhances privacy by allowing Alexa to distinguish between different users' voices. Additionally, users have control over their voice recordings and can delete them at any time.

However, ongoing discussions continue regarding data collection, storage, and the potential for inadvertent recordings. Striking the right balance between user convenience and privacy remains a challenge.

Challenges and Future Developments

As Alexa continues to evolve, several challenges and opportunities lie ahead:

1. **Privacy and Data Security:** Maintaining robust privacy protections and data security will be essential as voice assistants become more integrated into our lives.

2. **Personalization and Contextual Awareness:** Improving Alexa's ability to understand context and provide more personalized responses will enhance user experiences.

3. **Multimodal Interaction:** Expanding beyond voice to incorporate visual and tactile interactions will make Alexa more versatile.

4. **Localization:** Extending support for more languages and regions will broaden Alexa's global reach.

5. **AI Advancements:** Leveraging advancements in AI, machine learning, and natural language understanding will enable Alexa to offer even more sophisticated interactions.

Conclusion

Alexa, Amazon's virtual assistant, has left an indelible mark on modern society, redefining how we interact with technology and manage our daily lives. Its functions encompass a wide spectrum of tasks, from answering questions and controlling smart homes to providing entertainment and enhancing productivity.

As Alexa continues to advance and adapt to user needs, it remains a powerful symbol of the voice-first era, revolutionizing how we access information, connect with devices, and navigate our increasingly interconnected world. Alexa's journey from a nascent idea to a ubiquitous virtual assistant underscores the transformative potential of AI-driven technology in shaping our digital future.

Google Assistant: Google's AI Companion

Google Assistant: Google's AI Companion Shaping the Future

In an era of rapid technological advancement, virtual assistants have emerged as our trusted companions, simplifying daily tasks and providing information at our fingertips. Among the leading virtual assistants is Google Assistant, developed by tech giant Google. In this exploration, we will delve into the world of Google Assistant, tracing its evolution, functions, and the profound impact it has had on our daily lives.

The Birth of Google Assistant

Google Assistant is the culmination of Google's deep expertise in search, artificial intelligence, and natural language processing. Its origins can be traced back to the launch of Google Now, a contextual assistant introduced in 2012. Google Now provided users with proactive information based on their preferences and activities, setting the stage for more advanced virtual assistants.

In 2016, Google unveiled Google Assistant as a part of its Pixel smartphones and the Google Home smart speaker. This marked the official launch of a more conversational, AI-driven virtual assistant that would soon become available on a wide range of devices and platforms.

Understanding Google Assistant's Functions

Google Assistant is celebrated for its versatility, offering a broad spectrum of functions and capabilities that make it a valuable addition to our digital lives. Let's explore some of the key functions that Google Assistant provides:

1. Voice Commands and Natural Language Processing:

At the core of Google Assistant is voice interaction. Users can activate Google Assistant by saying "Hey Google" or "Okay Google," followed by a command or question. Google Assistant leverages advanced natural language processing to understand spoken language and respond in a conversational manner, enhancing the user experience.

2. Information Retrieval:

One of Google Assistant's primary functions is providing information and answering questions. Users can ask Google Assistant about a wide range of topics, from general knowledge queries and weather updates to sports scores and historical facts. Google Assistant fetches answers from the vast knowledge base of the internet and presents concise responses.

3. Smart Home Control:

Google Assistant serves as a central hub for smart home devices, enabling users to control lights, thermostats, locks, cameras, and more. Its compatibility with various smart home ecosystems, including Google's own Nest ecosystem and third-party devices, makes it a powerful tool for home automation.

4. Entertainment and Media:

Google Assistant enhances entertainment experiences by controlling music playback, providing recommendations, and connecting to streaming services. Users can ask Google Assistant to play music, read audiobooks, and even offer updates on movies, TV shows, and sports events.

5. Productivity and Organization:

Google Assistant simplifies daily routines by setting alarms, timers, and reminders. It integrates seamlessly with Google Calendar and other productivity apps, helping users manage appointments, schedules, and to-do lists.

6. Navigation and Directions:

Google Assistant's integration with Google Maps allows users to ask for directions, find nearby businesses, and check traffic conditions. It provides turn-by-turn navigation for driving, walking, and biking.

7. Third-Party Integrations:

One of Google Assistant's standout features is its extensive ecosystem of third-party integrations and apps. Developers can create actions that extend Google Assistant's functionality, enabling users to order food, book rides, play games, and perform a myriad of other tasks.

8. Multilingual Support:

Google Assistant supports multiple languages and dialects, making it accessible to users around the world. Its language capabilities continue to expand, catering to diverse linguistic needs.

9. Accessibility Features:

Google Assistant includes accessibility features, such as Voice Access, which allows users with mobility impairments to control their devices through voice commands. These features promote inclusivity and accessibility for users with disabilities.

The Impact of Google Assistant on Modern Society

Google Assistant's introduction into modern society has been transformative, influencing how we interact with technology and the ways we integrate it into our daily lives:

1. Redefining Digital Interaction:

Google Assistant is at the forefront of redefining how we interact with technology. Its conversational and natural language capabilities make it feel like a trusted companion, simplifying complex tasks and providing instant access to information.

2. Smart Home Revolution:

Google Assistant has played a pivotal role in popularizing the concept of the smart home. Its compatibility with a wide range of smart devices and ecosystems has made home automation more accessible and user-friendly.

3. Voice-First Era:

Google Assistant is a prominent symbol of the voice-first era, where voice commands replace traditional touch and typing interactions. This shift has implications for user interfaces and the way we interact with technology.

4. Ecosystem Expansion:

Google Assistant has expanded beyond smartphones and smart speakers to include a diverse range of devices, from smart displays and headphones to wearables and cars. This ecosystem provides users with a consistent and interconnected experience.

5. Multimodal Interaction:

Google Assistant's evolution includes multimodal interaction, incorporating visual and touch-based inputs alongside voice. This versatility makes it adaptable to various device form factors.

6. AI Advancements:

Leveraging Google's expertise in artificial intelligence, machine learning, and natural language understanding, Google Assistant continues to improve its ability to understand context and provide more personalized responses.

Ethical Considerations and Privacy

The widespread use of virtual assistants like Google Assistant raises ethical concerns, particularly regarding user privacy and data security. Google has implemented features to address these concerns, such as voice recognition and the ability to control and delete voice recordings. Users also have granular control over data sharing preferences.

However, discussions about data collection, storage, and user consent continue, as ensuring the right balance between user convenience and privacy remains a challenge.

Challenges and Future Developments

As Google Assistant continues to evolve, several challenges and opportunities lie ahead:

1. **Privacy and Data Security:** Maintaining robust privacy protections and data security will be essential as virtual assistants become more deeply integrated into our lives.

2. **Contextual Understanding:** Improving Google Assistant's ability to understand context and provide more nuanced and relevant responses will be a key focus for development.

3. **Global Localization:** Expanding support for more languages and regions will ensure that Google Assistant remains

accessible to users around the world.

4. **Seamless Multimodal Interaction:** Advancing multi-modal interactions to seamlessly transition between voice, touch, and visual inputs will enhance user experiences.

5. **Inclusivity:** Continuously improving accessibility features to ensure that Google Assistant is inclusive and accommodating for users with disabilities.

Conclusion

Google Assistant, Google's virtual companion, has left an indelible mark on modern society, redefining how we interact with technology and manage our daily lives. Its functions encompass a wide spectrum of tasks, from answering questions and controlling smart homes to providing entertainment and enhancing productivity.

As Google Assistant continues to advance and adapt to user needs, it remains a powerful symbol of the voice-first era, revolutionizing how we access information, connect with devices, and navigate our increasingly interconnected world. Google Assistant's journey from a visionary concept to an integral part of our digital lives underscores the transformative potential of AI-driven technology in shaping our digital future.

Cortana: Microsoft's Virtual Assistant

C ortana: Microsoft's Virtual Assistant Redefining Digital Interaction

In the realm of digital assistants, Cortana stands as Microsoft's contribution to the world of artificial intelligence and voice-driven interaction. Named after a beloved AI character from the Halo video game series, Cortana has evolved into a versatile and capable virtual assistant. In this exploration, we will delve into the world of Cortana, tracing its evolution, functions, and the profound impact it has had on the way we interact with technology.

The Origins of Cortana

Cortana's origins can be traced back to Microsoft's ambition to create a voice-activated, AI-driven virtual assistant. Microsoft announced Cortana's development in 2013, and it officially made its debut with the release of Windows Phone 8.1 in April 2014. The name "Cortana" was inspired by the character of the same name from the popular Halo video game series, known for its advanced AI capabilities and assistance to the game's protagonist, Master Chief.

Understanding Cortana's Functions

Cortana is celebrated for its wide range of functions and capabilities that simplify tasks and enhance productivity. Let's explore some of the key functions that Cortana provides:

1. Voice Commands and Natural Language Processing:

At its core, Cortana relies on voice interaction. Users can activate Cortana by saying, "Hey Cortana," followed by a command or question. Cortana leverages advanced natural language processing (NLP) to understand and respond to spoken language, creating a seamless and intuitive user experience.

2. Information Retrieval:

Cortana excels at providing information and answering questions. Users can ask Cortana about a myriad of topics, from general knowledge inquiries and weather updates to sports scores and historical facts. Cortana fetches answers from the web and presents concise responses.

3. Personal Productivity:

Cortana is a powerful personal assistant, capable of managing schedules, setting reminders, and creating to-do lists. It integrates seamlessly with Microsoft Office applications like Outlook, enabling users to schedule meetings and manage emails through voice commands.

4. Smart Home Control:

Cortana extends its functionality to control smart home devices, allowing users to manage lights, thermostats, locks, and other connected appliances. Its compatibility with various smart home ecosystems enhances home automation.

5. Entertainment and Media:

Cortana enhances entertainment experiences by controlling music playback, identifying songs, and providing movie recommendations. Users can ask Cortana to play music, read audiobooks, and receive updates on the latest movies and TV shows.

6. Navigation and Directions:

Cortana's integration with navigation services allows users to ask for directions, find nearby points of interest, and check traffic conditions. It provides turn-by-turn navigation for driving, walking, and biking.

7. Integration with Windows Ecosystem:

Cortana is deeply integrated into the Windows operating system, offering features like voice search, system settings control, and hands-free interaction with Windows devices.

8. Voice Dictation:

Cortana offers voice dictation capabilities, allowing users to dictate text for documents, emails, and messages with accuracy and convenience.

9. Accessibility Features:

Cortana includes accessibility features, such as screen reading capabilities, voice control for users with mobility impairments, and compatibility with assistive technologies. These features enhance accessibility and inclusivity.

The Impact of Cortana on Modern Society

Cortana's introduction into modern society has been transformative, reshaping the way we interact with technology and integrating AI-driven assistance into our daily lives:

1. Digital Interaction Evolution:

Cortana represents a pivotal step in the evolution of digital interaction. Its conversational AI capabilities have made technology more accessible and user-friendly.

2. Personal Productivity Enhancement:

Cortana has become an indispensable tool for personal productivity, helping users manage tasks, schedules, and communication more efficiently.

3. Integration into Windows Ecosystem:

Cortana's deep integration into the Windows ecosystem has created a seamless and interconnected user experience across devices, from PCs and laptops to tablets and smartphones.

4. Accessibility Advancements:

Cortana's accessibility features have made technology more inclusive, ensuring that users with disabilities can benefit from voice-driven interactions and assistive capabilities.

Challenges and Future Developments

As Cortana continues to evolve, several challenges and opportunities lie ahead:

1. **Competition in the Virtual Assistant Space:** Cortana faces stiff competition from other virtual assistants like Google Assistant, Amazon Alexa, and Apple's Siri. Continuously improving its capabilities and expanding its ecosystem will be key to maintaining relevance.

2. **Privacy and Data Security:** Ensuring robust privacy protections and data security features will be essential as virtual assistants handle increasingly sensitive information.

3. **Contextual Understanding:** Improving Cortana's ability to understand context and provide more nuanced and relevant responses will enhance user experiences.

4. **Global Localization:** Expanding support for more languages and regions will ensure that Cortana remains accessible to users around the world.

5. **Seamless Multimodal Interaction:** Advancing multimodal interactions to seamlessly transition between voice, touch, and visual inputs will enhance user experiences.

Conclusion

Cortana, Microsoft's virtual assistant, has left an indelible mark on modern society, redefining how we interact with technology and manage our daily lives. Its functions encompass a wide spectrum of tasks, from answering questions and controlling smart homes to enhancing personal productivity and entertainment experiences.

As Cortana continues to advance and adapt to user needs, it remains a powerful symbol of the growing role of AI-driven technology in shaping our digital future. Cortana's journey from conceptualization to widespread adoption underscores the transformative potential of virtual assistants in simplifying complex tasks and enhancing the way we connect with our devices and the digital world.

Bixby: Samsung's AI Voice Assistant

Bixby: Samsung's AI Voice Assistant Revolutionizing Digital Interaction

In the ever-evolving landscape of technology and artificial intelligence, Bixby emerges as Samsung's voice-driven AI assistant, redefining how we interact with our devices and the digital world. In this exploration, we will delve into the world of Bixby, tracing its evolution, functions, and the significant impact it has had on the user experience within the Samsung ecosystem.

The Genesis of Bixby

Bixby had its origins in Samsung's ambition to create an intelligent assistant that could seamlessly integrate with its range of consumer electronics. It was officially introduced in March 2017 alongside the launch of the Samsung Galaxy S8 smartphone. Unlike some virtual assistants that existed primarily on smartphones, Bixby was designed to extend its capabilities across a multitude of Samsung devices, from smartphones and tablets to smart refrigerators and televisions.

Understanding Bixby's Functions

Bixby stands out for its diverse set of functions and capabilities, all aimed at simplifying tasks, enhancing productivity, and providing a more intuitive user experience. Let's explore some of the key functions that Bixby provides:

1. Voice Commands and Natural Language Processing:

At its core, Bixby relies on voice interaction. Users can activate Bixby by saying "Hi Bixby" or by pressing a dedicated button on Samsung devices. Bixby leverages advanced natural language processing (NLP) to understand spoken language and respond in a conversational manner, making it more user-friendly.

2. Information Retrieval:

Bixby excels at providing information and answering questions. Users can ask Bixby about a wide range of topics, from general knowledge queries to weather forecasts and sports scores. Bixby fetches answers from the web and presents concise responses.

3. Personal Productivity:

Bixby serves as a personal assistant, helping users manage schedules, set reminders, and create to-do lists. It integrates seamlessly with Samsung's native applications, such as Samsung Notes and Calendar, streamlining tasks.

4. Device Control:

One of Bixby's unique features is its ability to control and interact with Samsung devices. Users can adjust settings, control smart home devices, and navigate menus using voice commands, creating a unified and interconnected ecosystem.

5. Visual Search and Translation:

Bixby incorporates visual search capabilities, enabling users to point their smartphone camera at objects or text and receive information about them. It also offers real-time translation, making it a valuable tool for travelers.

6. Integration with Samsung Ecosystem:

Bixby is deeply integrated into the Samsung ecosystem, offering features like device synchronization, where actions can seamlessly transition between Samsung devices, creating a consistent and interconnected user experience.

7. Third-Party Integrations:

While initially focused on Samsung's native applications and devices, Bixby has expanded its capabilities through third-party integrations, allowing users to interact with a broader range of services and applications.

8. Language Support:

Bixby supports multiple languages and dialects, making it accessible to users around the world. Its language capabilities continue to expand, catering to diverse linguistic needs.

The Impact of Bixby on Modern Society

Bixby's introduction into modern society has been transformative, influencing how users interact with technology, particularly within the Samsung ecosystem:

1. Unified Samsung Experience:

Bixby plays a pivotal role in creating a unified experience across Samsung devices. Users can seamlessly transition from their smartphone to their smart TV, refrigerator, or tablet, with Bixby serving as a bridge that connects these devices.

2. Enhanced Device Control:

Bixby's ability to control and interact with Samsung devices, including smart home appliances, has streamlined daily routines, making it easier for users to manage their connected environments.

3. Personalized Assistance:

Bixby's integration with native Samsung applications and services allows for personalized assistance and recommendations, enhancing productivity and user satisfaction.

Challenges and Future Developments

As Bixby continues to evolve, several challenges and opportunities lie ahead:

1. **Competition in the Virtual Assistant Space:** Bixby faces competition from established virtual assistants like Google Assistant, Amazon Alexa, and Apple's Siri. Continuously improving its capabilities and expanding third-party integrations will be key to its success.

2. **Global Localization:** Expanding support for more languages and regions will ensure that Bixby remains accessible and relevant to a global audience.

3. **Privacy and Data Security:** Ensuring robust privacy protections and data security features will be essential as Bixby handles increasingly sensitive information.

4. **Contextual Understanding:** Improving Bixby's ability to understand context and provide more nuanced and relevant responses will enhance user experiences.

Conclusion

Bixby, Samsung's AI voice assistant, has left an indelible mark on modern society, redefining how users interact with technology and manage their interconnected devices. Its functions encompass a wide range of tasks, from answering questions and controlling devices to enhancing productivity and providing a personalized user experience.

As Bixby continues to advance and adapt to user needs, it remains a symbol of Samsung's commitment to creating an interconnected ecosystem that simplifies daily tasks and enhances the way users connect with their devices and the digital world. Bixby's journey from concept to widespread adoption underscores the transformative potential of virtual assistants in shaping the future of technology and user interaction.

Other Virtual Assistants

Exploring Other Virtual Assistants: A Multifaceted Landscape

While Siri, Alexa, Google Assistant, Cortana, and Bixby are among the most well-known virtual assistants, they are not the only players in this rapidly evolving field. A diverse array of virtual assistants has emerged, each offering unique capabilities and catering to different niches and industries. In this exploration, we will delve into some of the other virtual assistants, their functions, and their contributions to modern society.

**1. IBM Watson:

IBM Watson is a cognitive AI system developed by IBM. It is known for its advanced natural language processing and machine learning capabilities. Watson's primary focus is on providing businesses and industries with AI-powered insights and solutions.

Functions:

- **Data Analysis:** Watson can analyze vast amounts of data, making it valuable for industries like healthcare, finance, and research.

- **Language Processing:** It understands and responds to natural language, making it useful for chatbots and customer support.

- **Healthcare Applications:** Watson is used in medical research, drug discovery, and patient diagnosis.

Impact:

IBM Watson has revolutionized industries by providing AI-powered insights and solutions. Its impact is most notable in healthcare, where it aids in diagnosis and treatment planning, and in finance, where it assists in fraud detection and risk assessment.

**2. Nuance Dragon:

Nuance Dragon is a virtual assistant that specializes in voice recognition and dictation. It is widely used in industries where accurate voice transcription and dictation are critical, such as healthcare and legal.

Functions:

- **Voice Recognition:** Nuance Dragon boasts high accuracy in transcribing spoken words into text.

- **Dictation:** It enables users to dictate documents and emails with precision.

- **Medical Transcription:** In healthcare, Nuance Dragon is used for medical transcription and documentation.

Impact:

Nuance Dragon has significantly improved efficiency in industries where voice transcription is essential. It has expedited documentation processes in healthcare, legal, and business contexts.

**3. Viv:

Viv is an AI platform known for its advanced conversational capabilities and contextual understanding. Samsung acquired Viv Labs in 2016, incorporating Viv's technology into its Bixby virtual assistant.

Functions:

- **Conversational AI:** Viv specializes in conversational interfaces, facilitating natural and context-aware interactions.

- **Third-Party Integration:** It can integrate with a wide range of third-party services and applications.

- **Personalization:** Viv can provide highly personalized recommendations and responses.

Impact:

While Viv's standalone presence is limited, its technology has influenced the capabilities of Samsung's Bixby, enhancing its conversational and third-party integration features.

4. Xiaoice:

Xiaoice is a virtual assistant developed by Microsoft specifically for Chinese-speaking users. It gained immense popularity in China for its conversational abilities and emotional intelligence.

Functions:

- **Conversational AI:** Xiaoice engages in natural, human-like conversations with users.

- **Emotional Understanding:** It can detect users' emotional states and respond accordingly.

- **Content Generation:** Xiaoice has been used to write poetry, stories, and even news articles.

Impact:

Xiaoice has had a significant cultural impact in China, where millions of users have engaged with it in conversations, turning to it for companionship and entertainment.

5. AliGenie:

AliGenie is Alibaba's virtual assistant designed for the Chinese market. It integrates with Alibaba's e-commerce platform and a wide range of smart devices.

Functions:

- **E-commerce Integration:** AliGenie allows users to shop on Alibaba's e-commerce platforms using voice commands.

- **Smart Home Control:** It can control smart devices in the home, from lights to appliances.

- **Content Delivery:** AliGenie provides news, weather updates, and other content.

Impact:

AliGenie's integration with Alibaba's e-commerce ecosystem has transformed the way Chinese consumers shop online, making it more convenient and accessible.

6. Mycroft:

Mycroft is an open-source virtual assistant that differentiates itself by its commitment to privacy and customization. It can be installed on various platforms and devices.

Functions:

- **Open Source:** Mycroft is open source, allowing developers to customize and extend its functionality.

- **Privacy-Focused:** It emphasizes user privacy and data ownership.

- **Voice Control:** Mycroft offers voice control for various applications and services.

Impact:

Mycroft's open-source nature and privacy focus have attracted users and developers who are concerned about data privacy and customization.

7. Hound:

Hound is a virtual assistant developed by SoundHound Inc. It is known for its speed and ability to process complex voice queries quickly.

Functions:

- **Fast Responses:** Hound provides rapid responses to voice queries, making it efficient for tasks like finding restaurants and making reservations.

- **Complex Queries:** It can understand and process multi-part queries and provide relevant results.

Impact:

Hound's speed and efficiency in handling complex voice queries have made it a preferred choice for tasks that require quick and accurate responses.

8. Lyra:

Lyra is an AI-powered virtual assistant developed by OpenAI. It is designed to facilitate natural language conversation and is available for developers to integrate into their applications.

Functions:

- **Natural Language Processing:** Lyra can understand and generate human-like text in multiple languages.

- **Conversational AI:** It engages in dialogues with users, making it suitable for chatbots and customer support.

Impact:

Lyra's natural language processing capabilities make it a valuable tool for developers looking to incorporate conversational AI into their applications.

9. Jasper:

Jasper is a virtual assistant developed by Facebook specifically for use within the workplace. It is designed to enhance productivity and streamline communication within organizations.

Functions:

- **Workplace Integration:** Jasper integrates with workplace collaboration tools like Slack and Microsoft Teams.

- **Task Automation:** It can automate tasks like scheduling meetings and fetching information from databases.

Impact:

Jasper aims to improve workplace efficiency by simplifying routine tasks and facilitating communication within teams.

10. Snips:

Snips is an AI voice assistant known for its focus on privacy and offline capabilities. It is designed to function without relying on cloud-based services.

Functions:

- **Privacy-Focused:** Snips processes voice commands locally on the device, minimizing data exposure.

- **Customization:** It allows users to create custom voice commands and applications.

Impact:

Snips' emphasis on privacy and offline capabilities has attracted users who prioritize data security and customization.

Conclusion

The landscape of virtual assistants is vast and diverse, with each offering unique functions and catering to specific niches and industries. These virtual assistants have played a pivotal role in redefining how we interact with technology, whether it's for personal productivity, entertainment, or industry-specific applications.

As virtual assistants continue to evolve and expand their capabilities, their influence on modern society will only grow. They represent a future where seamless, natural language interactions with technology are not only possible but also integral to our daily lives, transforming the way we work, communicate, and navigate our increasingly interconnected world.

How Virtual Assistants Work: The Power of Natural Language Processing (NLP)

The Power of Natural Language Processing (NLP)

Virtual assistants like Siri, Alexa, Google Assistant, Cortana, Bixby, and others have become integral parts of our digital lives. They simplify tasks, provide information, and enable natural, voice-driven interactions with technology. At the heart of these virtual assistants lies Natural Language Processing (NLP), a sophisticated technology that empowers them to understand, interpret, and respond to human language. In this exploration, we will delve into the workings of virtual assistants, emphasizing the pivotal role of NLP.

1. Voice Input:

The interaction with a virtual assistant typically begins with a voice command or query spoken by the user. For instance, saying "Hey Siri" or "Okay Google" activates the assistant, signaling that the user is ready to ask a question or give a command.

2. Speech Recognition:

The virtual assistant's first task is to transcribe the spoken words into text. This process is known as speech recognition. Advanced algorithms and models analyze the audio input and convert it into written words. Speech recognition systems have evolved significantly, achieving high accuracy rates even in noisy environments.

3. Natural Language Understanding (NLU):

Once the spoken words are transcribed into text, the NLU component of the virtual assistant takes over. NLU is a critical aspect of NLP that focuses on comprehending the meaning and intent behind the words. It breaks down the text into key components, such as:

- **Intent:** What action or task the user wants to perform (e.g., checking the weather, setting an alarm).

- **Entities:** Specific pieces of information required to complete the task (e.g., the location for weather, the time for an alarm).

- **Context:** The context of the conversation, including previous commands or questions.

NLU employs machine learning models, neural networks, and statistical analysis to understand the user's intent accurately.

4. Query Processing:

Once the virtual assistant has a clear understanding of the user's intent and the required information, it processes the query. This may involve retrieving data from the internet, accessing the device's local resources, or connecting to external services and APIs to gather relevant information.

5. Response Generation:

After processing the query, the virtual assistant generates a response. This response can take various forms, such as:

- **Spoken Response:** In cases where the interaction is voice-driven, the assistant may generate a spoken reply.

- **Text Response:** If the interaction is text-based (e.g., chatbots or text-based assistants), the response is in written form.

- **Action or Task:** For certain commands, the virtual assistant may execute an action or task directly, such as setting an alarm or sending a message.

6. Voice Synthesis (Text-to-Speech):

If the response is intended to be spoken, the virtual assistant employs Text-to-Speech (TTS) technology to convert the written text into natural-sounding speech. TTS algorithms strive to replicate human speech patterns, including intonation and pacing.

7. Presentation to User:

The final step is presenting the response to the user. This can occur through various devices and interfaces, such as smartphones, smart speakers, smart displays, or chat windows on a computer. The user receives the information or action taken by the virtual assistant.

8. **Continuous Learning and Improvement:**

Virtual assistants are designed to continuously learn and improve their performance. They leverage machine learning techniques to adapt to user preferences, understand context better, and provide increasingly accurate responses over time. This learning process involves analyzing user interactions and feedback.

Challenges and Future Developments

While virtual assistants have made remarkable strides, challenges persist:

- **Privacy and Data Security:** Safeguarding user data and ensuring that personal information is handled securely remains a paramount concern.

- **Multimodal Interaction:** Future virtual assistants will excel at multimodal interactions, combining voice, text, and visual inputs seamlessly.

- **Contextual Understanding:** Enhancing the ability to understand context, carry on multi-turn conversations, and provide more accurate responses is a key focus.

- **Language Support:** Expanding support for multiple languages and dialects to cater to diverse linguistic needs is crucial for global accessibility.

- **Ethical Considerations:** Ensuring ethical use of virtual assistants and addressing issues like bias in AI responses will be ongoing concerns.

Conclusion

Virtual assistants have transformed the way we interact with technology, making it more natural and intuitive. Central to their functionality is Natural Language Processing (NLP), which enables them to understand, interpret, and respond to human language effectively. As NLP and AI technologies continue to advance, virtual assistants will become even more integral to our daily lives, streamlining tasks, providing information, and offering assistance across various devices and platforms. The future holds exciting possibilities for enhanced communication and collaboration between humans and AI-powered virtual assistants.

How Speech Recognition Works

Speech recognition, also known as Automatic Speech Recognition (ASR) or voice recognition, is a technology that allows a computer or machine to convert spoken language into written text. It's a critical component of many applications, including virtual assistants, transcription services, voice commands for devices, and more. Let's delve into how speech recognition works and its significance in modern technology.

How Speech Recognition Works:

1. **Audio Input:** The process begins with an audio input, which is typically spoken language captured by a microphone. This can be a user's voice command to a virtual assistant, a recorded conversation, or any form of spoken communication.

2. **Audio Preprocessing:** The audio input often undergoes preprocessing to enhance its quality and reduce noise. This

can involve filtering out background noise, normalizing audio levels, and removing echoes to improve the accuracy of recognition.

3. **Feature Extraction:** The next step involves extracting relevant features from the audio signal. These features can include spectral characteristics like the frequencies and amplitudes of the sound waves over time. Common techniques include Mel-Frequency Cepstral Coefficients (MFCCs) and Linear Predictive Coding (LPC).

4. **Acoustic Modeling:** Acoustic modeling is a crucial component of speech recognition. It involves building statistical models that relate the extracted audio features to phonemes or smaller units of speech sound. Phonemes are the fundamental sound units of a language. The models are trained using extensive datasets of recorded speech and associated transcriptions.

5. **Language Modeling:** In addition to understanding the acoustic characteristics of speech, speech recognition systems need to consider the language context. Language models predict the likelihood of a particular word or sequence of words occurring in a given language. N-gram models and neural language models are commonly used for this purpose.

6. **Decoding:** Once the acoustic and language models are trained and ready, the recognition system uses a decoding process to determine the most likely transcription of the spoken words. This process involves matching the audio features to the models and generating a ranked list of possible

word sequences.

7. **Post-processing:** In this stage, post-processing techniques may be applied to refine the transcription. These can include grammar checking, spell checking, and contextual analysis to improve accuracy and correct errors.

8. **Output Text:** The final result of the speech recognition process is a written transcript of the spoken language. This transcript can be used for various purposes, from transcribing spoken audio to generating text-based commands for devices or applications.

Significance of Speech Recognition:

1. **Accessibility:** Speech recognition technology has greatly improved accessibility for individuals with disabilities. It enables voice-controlled devices and software, making technology more inclusive.

2. **Convenience:** Speech recognition simplifies human-computer interaction. Users can dictate text, give voice commands to devices, and perform tasks without the need for manual input.

3. **Productivity:** In business and healthcare, speech recognition tools expedite documentation processes. Professionals can dictate notes, reports, and records more efficiently, reducing paperwork.

4. **Virtual Assistants:** Speech recognition is a core component of virtual assistants like Siri, Google Assistant, and Alexa. It enables these assistants to understand and respond to user

voice commands and queries.

5. **Transcription Services:** Transcription services rely on speech recognition to convert audio and video content into written text quickly and accurately.

6. **Automotive Technology:** In-car voice recognition systems allow drivers to control navigation, entertainment, and communication without taking their hands off the wheel.

7. **Language Translation:** Speech recognition technology can be used for real-time language translation, breaking down language barriers in communication.

8. **Security:** Voice recognition is employed in security systems for voice authentication, adding an extra layer of identity verification.

9. **Search Engines:** Speech recognition is used in voice search on search engines, enabling users to find information by speaking their queries.

10. **Customer Support:** Many companies use speech recognition for interactive voice response (IVR) systems to handle customer inquiries efficiently.

Speech recognition has come a long way, thanks to advances in machine learning and neural networks. Its continued development holds promise for even more accurate and versatile applications across various industries, making voice-driven interaction with technology more seamless and intuitive.

Machine Learning and Training

M achine learning (ML) and training are fundamental components of artificial intelligence (AI) systems, including virtual assistants and speech recognition. Let's explore what machine learning is, how it works, and the training process that powers these technologies.

Machine Learning (ML):

Machine learning is a subfield of artificial intelligence that focuses on the development of algorithms and models that enable computer systems to learn from and make predictions or decisions based on data. ML systems improve their performance through experience, rather than relying on explicit programming.

Key Concepts in Machine Learning:

1. **Data:** Data is the lifeblood of machine learning. ML algorithms require large datasets to learn patterns and make predictions. In the context of virtual assistants and speech recognition, data includes recorded speech, text transcrip-

tions, and user interactions.

2. **Features:** Features are the characteristics or attributes extracted from data that the ML algorithm uses for learning. In speech recognition, features might include acoustic characteristics like frequency and amplitude.

3. **Models:** ML models are mathematical representations of the relationships between data features and the desired output. For speech recognition, models map audio features to text transcriptions.

4. **Training:** Training is the process by which ML models learn from data. It involves exposing the model to labeled examples (input-output pairs) and adjusting its internal parameters to minimize prediction errors.

5. **Testing and Evaluation:** After training, models are tested on new, unseen data to assess their performance. Common evaluation metrics include accuracy, precision, recall, and F1-score.

Machine Learning Process:

1. **Data Collection:** The first step in ML is collecting and preparing data. In speech recognition, this involves recording spoken language samples and creating transcriptions.

2. **Feature Extraction:** Relevant features are extracted from the data. For speech recognition, this includes acoustic features like MFCCs or spectrograms.

3. **Model Selection:** Researchers and engineers choose an appropriate ML model architecture based on the problem at

hand. For speech recognition, deep neural networks (DNNs) and recurrent neural networks (RNNs) are commonly used.

4. **Training:** The selected model is trained using labeled data. During training, the model learns to map input features (e.g., audio) to target labels (e.g., transcriptions) by adjusting its internal parameters through optimization techniques like gradient descent.

5. **Validation and Hyperparameter Tuning:** To fine-tune the model's performance, hyperparameters are adjusted, and the model's performance is validated on a separate validation dataset.

6. **Testing:** The trained model is evaluated on a test dataset to assess its real-world performance.

7. **Deployment:** Once the model meets performance criteria, it can be deployed in applications like virtual assistants or speech recognition systems.

Training in Virtual Assistants and Speech Recognition:

In the context of virtual assistants and speech recognition, training involves large datasets of recorded speech and corresponding text transcriptions. These datasets are used to train acoustic models (for speech recognition) and natural language understanding models (for virtual assistants). Training such models requires significant computational resources and can be done using specialized hardware like graphics processing units (GPUs) or TPUs (Tensor Processing Units).

The training process typically involves iterations, where models are trained, evaluated, and adjusted multiple times to improve performance. Researchers continually work on refining model architectures and training techniques to achieve higher accuracy and better under-standing of user input.

Overall, machine learning and training are at the core of modern speech recognition and virtual assistant systems, enabling them to understand spoken language, recognize patterns, and provide mean-ingful responses to users. Advances in these fields continue to drive improvements in AI-driven technologies, making them increasingly accurate and useful in various applications.

Cloud Computing and Data Processing

Enabling Scalable and Efficient Information Management

In the era of data-driven decision-making and digital transformation, cloud computing and data processing have emerged as critical technologies that power the modern business landscape. These technologies not only provide organizations with the infrastructure and tools to store and process vast amounts of data but also offer the flexibility and scalability required to adapt to rapidly changing business needs. In this comprehensive exploration, we will delve into the world of cloud computing and data processing, understanding their fundamental concepts, benefits, challenges, and real-world applications.

Understanding Cloud Computing

What Is Cloud Computing?

Cloud computing refers to the delivery of computing services, including servers, storage, databases, networking, software, analytics, and intelligence, over the internet (the cloud) to offer faster innovation, flexible resources, and economies of scale. Instead of owning their computing infrastructure or data centers, companies can rent access to anything from applications to storage from a cloud service provider.

Key Characteristics of Cloud Computing:

1. **On-Demand Self-Service:** Users can provision and manage resources as needed, without human intervention from the service provider.

2. **Broad Network Access:** Services are accessible over the internet via various devices, including laptops, smartphones, and tablets.

3. **Resource Pooling:** Resources are pooled and shared among multiple users, with each user's data and applications isolated from others.

4. **Rapid Elasticity:** Resources can be scaled up or down quickly to accommodate changing workloads, ensuring optimal resource utilization.

5. **Measured Service:** Cloud computing resources are metered, and users are billed based on their actual usage, providing cost-effective flexibility.

Cloud Service Models:

Cloud computing offers various service models to cater to different needs:

1. **Infrastructure as a Service (IaaS):** IaaS provides virtual-

ized computing resources over the internet, including virtual machines, storage, and networking. Users manage the operating systems, applications, and data.

2. **Platform as a Service (PaaS):** PaaS offers a platform and environment to develop, test, and deploy applications. Users don't need to worry about managing the underlying infrastructure.

3. **Software as a Service (SaaS):** SaaS delivers software applications over the internet on a subscription basis. Users access the software through a web browser, eliminating the need for installation and maintenance.

Benefits of Cloud Computing:

Cloud computing provides numerous advantages for organizations and individuals:

1. **Scalability:** Cloud resources can be scaled up or down as needed, ensuring that computing power matches demand, and minimizing costs.

2. **Cost-Efficiency:** Users pay only for the resources they use, eliminating the need for large upfront capital investments in infrastructure.

3. **Flexibility and Accessibility:** Cloud services can be accessed from anywhere with an internet connection, enabling remote work and global collaboration.

4. **Speed and Agility:** Cloud services can be provisioned and deployed rapidly, reducing time to market for applications and services.

5. **Reliability and Redundancy:** Cloud providers often offer high levels of uptime and data redundancy to ensure data availability.

Cloud Deployment Models:

Cloud computing can be deployed in various ways to meet specific needs:

1. **Public Cloud:** Services are provided by a third-party cloud service provider, and resources are shared among multiple organizations.

2. **Private Cloud:** Resources are exclusively used by a single organization, providing greater control and security but with higher infrastructure costs.

3. **Hybrid Cloud:** A combination of public and private clouds allows data and applications to be shared between them while maintaining some level of isolation.

4. **Multi-Cloud:** Organizations use services from multiple cloud providers to avoid vendor lock-in and leverage the strengths of each provider.

Understanding Data Processing

What Is Data Processing?

Data processing refers to the collection, manipulation, and transformation of data into meaningful information. It encompasses a wide range of activities, from simple data entry and storage to complex analytics and machine learning. Effective data processing is crucial for organizations to derive insights, make informed decisions, and gain a competitive edge.

Key Stages of Data Processing:

1. **Data Collection:** The process begins with the collection of raw data from various sources, including sensors, devices, databases, and external data feeds.

2. **Data Preparation:** Raw data often requires cleaning, transformation, and formatting to ensure its quality and compatibility with analysis tools.

3. **Data Analysis:** This stage involves using statistical and analytical techniques to uncover patterns, trends, and insights within the data.

4. **Data Visualization:** Visual representations like charts, graphs, and dashboards help communicate the findings effectively to stakeholders.

5. **Data Storage:** Processed data is stored in databases, data warehouses, or data lakes for easy retrieval and future analysis.

6. **Data Reporting:** Generated reports and summaries provide a structured view of the data, facilitating decision-making.

Challenges in Data Processing:
Effective data processing faces several challenges:

1. **Data Volume:** The exponential growth of data, often referred to as "big data," requires scalable solutions for storage and processing.

2. **Data Variety:** Data comes in various formats, including structured, semi-structured, and unstructured, making integration and processing complex.

3. **Data Velocity:** The speed at which data is generated and needs to be processed in real-time presents challenges for timely analysis.

4. **Data Quality:** Ensuring data accuracy and consistency is essential for meaningful analysis and decision-making.

Cloud Computing and Data Processing: A Powerful Duo

Cloud computing and data processing are deeply intertwined, with cloud services providing the ideal environment for efficient and scalable data processing. Here's how they work together:

1. **Scalable Resources:** Cloud platforms offer elastic resources, allowing organizations to scale up or down based on data processing requirements. This flexibility ensures optimal resource allocation and cost efficiency.

2. **Storage and Databases:** Cloud providers offer a range of storage and database services, such as Amazon S3, Azure Blob Storage, and Google Cloud Storage, which can handle vast amounts of data efficiently.

3. **Analytics Services:** Cloud providers offer specialized analytics services, like Amazon Redshift, Google BigQuery, and Azure Data Lake Analytics, that enable advanced data processing and analysis.

4. **Machine Learning and AI:** Cloud platforms provide access to machine learning and AI services, enabling organizations to build predictive models, automate data processing tasks, and gain deeper insights.

5. **Data Integration:** Cloud-based ETL (Extract, Transform,

Load) services simplify the integration of data from multiple sources, making it accessible for processing.

Real-World Applications

Cloud computing and data processing have transformed numerous industries and applications:

1. **E-commerce:** Online retailers analyze customer behavior and purchase patterns to personalize recommendations and optimize inventory management.

2. **Healthcare:** Electronic health records (EHRs) are processed in the cloud to improve patient care, research, and billing processes.

3. **Finance:** Banks use data processing and analytics to detect fraudulent transactions, assess credit risk, and make investment decisions.

4. **Manufacturing:** IoT devices collect data on machinery performance and product quality, enabling predictive maintenance and process optimization.

5. **Media and Entertainment:** Streaming platforms use data to recommend content, tailor advertisements, and analyze user engagement.

6. **Transportation and Logistics:** GPS data and route optimization algorithms improve delivery efficiency and reduce fuel consumption.

Challenges and Considerations

While cloud computing and data processing offer immense benefits, organizations should consider challenges such as data security, compliance with data protection regulations, and vendor lock-in. Ensuring data privacy and maintaining data integrity are critical concerns, especially when handling sensitive or personally identifiable information.

Conclusion

Cloud computing and data processing are the dynamic duo powering the digital transformation of businesses and organizations across various industries. The scalable and flexible nature of cloud services, combined with sophisticated data processing capabilities, enables efficient data-driven decision-making, innovation, and improved customer experiences.

As technology continues to evolve, cloud computing and data processing will play an increasingly pivotal role in harnessing the power of data to address complex challenges and uncover new opportunities in the ever-changing landscape of the digital age. Organizations that leverage these technologies effectively will be better positioned to thrive and adapt in an increasingly data-centric world.

User Experience Design (UX Design)

U ser Experience Design (UX Design) is a critical discipline within the field of technology and product development. It focuses on creating products and interfaces that provide meaningful and satisfying experiences for users. In this exploration, we will delve into the key principles, processes, and importance of user experience design.

Key Principles of User Experience Design:

1. **User-Centered Design:** UX design starts with understanding the needs, behaviors, and preferences of users. The design process revolves around creating solutions that align with user goals.

2. **Usability:** Products should be easy to use and navigate. Usability testing and iterative design are common methods to

ensure a user-friendly experience.

3. **Accessibility:** Designing for accessibility ensures that products are usable by individuals with disabilities. This includes considerations for screen readers, keyboard navigation, and alternative input methods.

4. **Consistency:** Maintaining consistency in the design elements, such as colors, fonts, and interactions, helps users understand and predict how a product works.

5. **Clarity:** Clear and concise communication, both in terms of content and interface elements, is crucial to guide users effectively.

6. **Feedback and Affordances:** Providing feedback to users about the outcome of their actions and making interactive elements visually suggestive of their functionality (affordances) enhance user understanding.

7. **Emotional Design:** Beyond functionality, design can evoke emotions. Creating a positive emotional connection with users can lead to stronger engagement and loyalty.

The UX Design Process:
The UX design process typically involves several stages:

1. **Research:** Understanding the target audience, their needs, pain points, and goals through user research, surveys, and interviews.

2. **Ideation:** Brainstorming and generating ideas for potential solutions, often involving techniques like sketching, wireframing, and prototyping.

3. **Prototyping:** Creating low-fidelity and high-fidelity prototypes to test and refine design concepts.

4. **Testing and Evaluation:** Conducting usability testing and gathering user feedback to identify issues and make improvements.

5. **Implementation:** Collaborating with developers and engineers to bring the design to life in a functional product.

6. **Iteration:** Continuously refining the design based on user feedback and evolving needs.

Importance of User Experience Design:

1. **User Satisfaction:** A well-designed user experience leads to satisfied users who are more likely to engage with a product or service.

2. **Reduced Learning Curve:** Intuitive and user-friendly designs reduce the time and effort users need to learn and use a product.

3. **Customer Loyalty:** A positive user experience fosters customer loyalty and can lead to repeat business and referrals.

4. **Competitive Advantage:** In crowded markets, superior user experience can set a product apart and give a competitive edge.

5. **Reduced Support Costs:** Good UX design can reduce the need for extensive customer support, saving time and resources.

UX Design in Different Domains:

1. **Web Design:** UX design plays a critical role in creating websites that are easy to navigate, visually appealing, and provide valuable content.

2. **Mobile App Design:** Mobile app success often hinges on an intuitive user interface and a seamless user experience.

3. **E-commerce:** In online shopping, UX design influences factors like product discovery, cart abandonment rates, and overall customer satisfaction.

4. **Software Development:** In software applications, UX design ensures that users can efficiently perform tasks and access features.

5. **Product Design:** For physical products, UX design considers not only aesthetics but also ergonomics and usability.

Challenges in UX Design:

1. **Balancing User Needs and Business Goals:** Designers often need to strike a balance between what users want and what aligns with business objectives.

2. **Continuous Adaptation:** User expectations and technology evolve, requiring designers to adapt and keep products up-to-date.

3. **Cross-Platform Consistency:** Maintaining a consistent user experience across different platforms and devices can be challenging.

4. **Accessibility:** Ensuring that products are accessible to users

with disabilities requires specific expertise and effort.

Emerging Trends in UX Design:

1. **Voice User Interfaces (VUI):** Designing interfaces for voice-activated systems like smart speakers and virtual assistants.

2. **Augmented and Virtual Reality (AR/VR):** Creating immersive and intuitive experiences in augmented and virtual environments.

3. **AI and Personalization:** Leveraging artificial intelligence to provide personalized user experiences and recommendations.

4. **Ethical Design:** Addressing ethical considerations in design, including issues of privacy, bias, and user well-being.

In conclusion, user experience design is a fundamental aspect of technology and product development. It not only influences the success of products but also shapes how users interact with technology in their daily lives. By prioritizing user-centered design principles and following a structured design process, organizations can create products and interfaces that not only meet user needs but also provide a competitive advantage in the market. As technology continues to evolve, UX design will play a central role in shaping the future of human-computer interaction.

Ethical considerations

E thical considerations are of paramount importance in all aspects of technology, including artificial intelligence, data processing, user experience design, and human-AI interactions. Ensuring ethical practices is vital to prevent harm, protect privacy, and maintain trust in these technologies. Let's explore some key ethical considerations in these domains:

1. Privacy and Data Protection:

- **Data Collection:** Organizations should collect only the data necessary for a specific purpose and obtain informed consent from users.

- **Data Storage:** Safeguarding data through encryption and access controls is crucial to prevent unauthorized access or breaches.

- **Data Sharing:** Transparent policies should govern data sharing and ensure that data is not misused or sold without user consent.

- **Data Retention:** Organizations should establish clear

guidelines for how long data is retained and under what conditions it is deleted.

2. Bias and Fairness:

- **Algorithmic Bias:** Developers should mitigate bias in AI algorithms to prevent discriminatory outcomes, especially in areas like hiring, lending, and criminal justice.

- **Fair Representation:** Diverse datasets and perspectives should be considered to avoid underrepresentation or marginalization of specific groups.

3. Transparency and Accountability:

- **Algorithm Transparency:** Organizations should disclose how algorithms make decisions to ensure transparency and accountability.

- **Accountability for Errors:** When AI systems make mistakes, organizations should take responsibility and provide avenues for redress.

4. Informed Consent:

- **User Consent:** Users should be fully informed about how their data will be used and have the ability to opt in or out of data collection and processing.

5. User-Centered Design:

- **Design Ethics:** Designers should prioritize user well-being, avoid dark patterns, and ensure that designs do not manipulate or deceive users.

6. Accessibility:

- **Inclusive Design:** Products and interfaces should be designed to be accessible to individuals with disabilities, ensuring equitable access to technology.

7. Ethical AI and Automation:

- **Human-in-the-Loop:** AI systems that automate decisions should include a human-in-the-loop for oversight and intervention when necessary.

8. Consent in Human-AI Interaction:

- **Clear Communication:** Virtual assistants and chatbots should inform users when they are interacting with AI and offer options for human assistance.

9. Ethical Research and Development:

- **Responsible Research:** Ethical considerations should guide research involving AI, including the potential societal impact of new technologies.

10. Ethical Guidelines and Codes of Conduct:

- **Industry Standards:** Organizations should adhere to industry-specific ethical guidelines and codes of conduct, such as the ACM Code of Ethics and Professional Conduct for computer professionals.

11. Ethical Use of Data:

- **Sensitive Data:** Organizations should handle sensitive data with care and ensure it is not used for harmful purposes or discriminatory practices.

12. Ethical Considerations in Emerging Technologies:

- **AR/VR and Ethics:** In augmented and virtual reality, designers should consider ethical issues related to user experiences, including privacy in virtual spaces.

13. Ethical AI Governance:

- **Ethical AI Committees:** Organizations should establish committees or boards to oversee the ethical use of AI and technology.

14. Ethical Decision-Making:

- **Ethical Frameworks:** Organizations should adopt ethical decision-making frameworks to guide complex decisions involving technology and data.

15. Continuous Monitoring and Improvement:

- **Ethics Audits:** Regular audits and assessments of technology and data practices can help organizations identify and address ethical issues.

16. Ethical Education and Training:

- **Ethics Training:** Organizations should provide ethics training to employees, developers, and other stakeholders to foster an ethical culture.

Ethical considerations are not static; they evolve alongside technology. Therefore, staying informed about emerging ethical challenges and adapting practices accordingly is essential. By prioritizing ethical principles, organizations can not only avoid harm but also build trust, foster innovation, and contribute to a more responsible and equitable technological future.

Privacy Concerns

P rivacy concerns are a central issue in the digital age, especially with the proliferation of technologies like artificial intelligence, data processing, and virtual assistants. These technologies have the potential to collect, store, and analyze vast amounts of personal information, raising significant privacy considerations. Let's explore some of the key privacy concerns in these domains:

1. Data Collection and Retention:

- **Concern:** Organizations often collect more data than necessary, and there is a lack of transparency about what data is being collected and how long it will be retained.

- **Impact:** Overcollection and indefinite retention of data can lead to privacy breaches, unauthorized access, and potential misuse of personal information.

2. Consent and Informed Choice:

- **Concern:** Users may not fully understand or have control over how their data is collected, used, and shared.

- **Impact:** Lack of informed consent can result in users unknowingly sharing sensitive information, leading to privacy violations.

3. Data Security:

- **Concern:** Inadequate security measures can lead to data breaches and unauthorized access to personal information.

- **Impact:** Stolen or compromised data can lead to identity theft, financial fraud, and other serious privacy and security risks.

4. Algorithmic Bias:

- **Concern:** AI and machine learning algorithms can inherit biases from training data, leading to discriminatory outcomes, particularly in areas like hiring and lending.

- **Impact:** Algorithmic bias can reinforce societal biases and lead to unfair and discriminatory practices.

5. Voice Data and Virtual Assistants:

- **Concern:** Virtual assistants like Siri and Alexa may inadvertently record and store private conversations without clear user consent.

- **Impact:** Unauthorized voice data retention can infringe on users' privacy and lead to concerns about surveillance.

6. Location Data:

- **Concern:** Mobile devices and apps often collect precise location data, raising concerns about user tracking.

- **Impact:** Unauthorized access to location data can be used for stalking or invasive advertising practices.

7. Facial Recognition:

- **Concern:** The use of facial recognition technology for surveillance and identification purposes can infringe on privacy rights.

- **Impact:** Widespread use of facial recognition can lead to mass surveillance, privacy violations, and misuse of biometric data.

8. Data Sharing and Third-Party Access:

- **Concern:** Organizations may share user data with third-party entities without users' explicit consent or knowledge.

- **Impact:** Data shared with third parties can be used for targeted advertising, profiling, or other purposes without user control.

9. De-Anonymization:

- **Concern:** Even anonymized data can often be de-anonymized when combined with other information, potentially revealing sensitive details about individuals.

- **Impact:** De-anonymization techniques can undermine privacy protections and expose individuals to risks.

10. Data Minimization:

- **Concern:** Some organizations may collect more data than necessary for a particular service or function.
- **Impact:** Overcollection of data can increase the risk of privacy breaches and make it challenging to maintain data privacy.

11. Consent for Children:

- **Concern:** Children may not fully understand the implications of sharing personal information online, and obtaining proper consent can be challenging.

- **Impact:** Inadequate safeguards for children's privacy can expose them to potential risks and exploitation.

Addressing these privacy concerns requires a combination of legal regulations, responsible data practices, and user education:

- **Privacy Regulations:** Governments around the world have enacted privacy regulations such as the General Data Protection Regulation (GDPR) in Europe and the California Consumer Privacy Act (CCPA) in the United States to protect individual privacy rights.

- **Ethical Data Practices:** Organizations should adopt ethical data collection and processing practices, including data minimization, clear consent mechanisms, and strong security measures.

- **User Education:** Users should be informed about their privacy rights and provided with clear, user-friendly privacy policies.

- **Transparency:** Organizations should be transparent about their data practices, including data collection, sharing, and retention policies.

Balancing the benefits of technology with individual privacy rights is a complex challenge. Striking the right balance involves ongoing efforts by individuals, organizations, policymakers, and technology developers to ensure that privacy concerns are addressed while still fostering innovation and technological progress.

Virtual Assistants: Transforming Lives in Smart Homes

Virtual Assistants: Transforming Lives in Smart Homes

Smart homes have become synonymous with convenience and automation, and at the heart of this transformation are virtual assistants. These intelligent digital companions, powered by artificial intelligence (AI), play a pivotal role in enhancing the functionality and comfort of modern homes.

1. The Rise of Smart Homes and Virtual Assistants:

The concept of a smart home revolves around the integration of various devices, sensors, and systems to automate and enhance daily tasks. Virtual assistants, such as Amazon's Alexa, Google Assistant, and Apple's Siri, have become essential components of these smart ecosystems.

2. Voice-Activated Control:

One of the key features of virtual assistants is voice-activated control. Users can interact with their smart devices simply by issuing voice commands, making it effortless to adjust lighting, control thermostats, play music, and even order groceries.

3. Home Automation:

Virtual assistants serve as the central hub for home automation. They connect with smart devices like thermostats, lights, locks, and cameras, enabling users to create custom routines and automate tasks based on their preferences.

4. Personalized Experiences:

Virtual assistants leverage machine learning to understand user preferences and habits over time. This enables them to provide personalized recommendations and perform actions tailored to the user's needs.

5. Safety and Security:

In addition to convenience, virtual assistants contribute to home security. They can integrate with security systems and cameras to provide real-time updates and allow users to monitor their homes remotely.

6. Challenges and Concerns:

While virtual assistants offer numerous benefits in smart homes, there are also privacy and security concerns. The always-listening nature of these devices can raise questions about data privacy, prompting the need for robust security measures and user education.

Virtual Assistants in Personal Devices: Redefining User Experiences

Personal devices like smartphones, tablets, and wearable gadgets have become indispensable in our daily lives. Virtual assistants integrated into these devices have transformed how we interact with technology and access information.

1. Mobile Assistants:

Virtual assistants on smartphones, such as Siri (Apple), Google Assistant (Android), and Bixby (Samsung), have become digital companions that assist users with a wide range of tasks.

2. Voice-Activated Assistance:

Voice-activated virtual assistants offer hands-free convenience. Users can send messages, set reminders, check the weather, and perform web searches, all by using voice commands.

3. Integration with Apps and Services:

Virtual assistants seamlessly integrate with apps and services, allowing users to perform tasks like booking rides, making restaurant reservations, or controlling smart home devices without leaving their preferred virtual assistant.

4. Accessibility Features:

Virtual assistants have made personal devices more accessible to individuals with disabilities. Features like voice commands and screen readers empower users with visual or motor impairments.

5. Predictive and Contextual Assistance:

AI-driven virtual assistants can predict user needs based on context. For instance, they might provide traffic updates when they know the user is about to leave for work.

6. Challenges and Privacy:

The growing integration of virtual assistants into personal devices raises concerns about data privacy and the potential for eavesdropping. Tech companies continually work to address these challenges by implementing privacy features and providing users with more control over their data.

Virtual Assistants in the Workplace: Boosting Productivity and Efficiency

Virtual assistants are not limited to homes and personal devices; they also play a significant role in the workplace. From scheduling meetings to performing research, they help streamline tasks and enhance productivity.

1. Office Productivity:

Virtual assistants like Microsoft's Cortana and Slack's integrations provide office workers with valuable tools for managing their schedules, setting reminders, and even conducting research.

2. Task Automation:

Repetitive administrative tasks can be automated by virtual assistants. This allows employees to focus on more value-added activities, increasing overall productivity.

3. Enhanced Collaboration:

Virtual assistants in collaboration tools like Slack and Microsoft Teams facilitate communication and information retrieval, making teamwork more efficient.

4. Personalized Assistance:

In the workplace, virtual assistants can provide personalized assistance to employees, helping them find relevant documents, schedule meetings, and access relevant data.

5. Data Insights:

Advanced analytics capabilities enable virtual assistants to extract insights from data, helping organizations make informed decisions and gain a competitive edge.

6. Security and Privacy:

In corporate environments, security and data privacy are paramount. Employers must ensure that virtual assistants adhere to strict security protocols to protect sensitive information.

Virtual Assistants in Healthcare: Revolutionizing Patient Care

Virtual assistants are making significant inroads into the healthcare sector, offering innovative solutions for patient care, healthcare providers, and medical research.

1. Medical Information Retrieval:

Virtual assistants can access medical databases and provide healthcare professionals with quick access to relevant medical information, research, and treatment guidelines.

2. Appointment Scheduling:

Patients can use virtual assistants to schedule doctor's appointments, reducing wait times and administrative burdens for healthcare providers.

3. Medication Management:

Virtual assistants help patients manage their medications by sending reminders, providing dosage instructions, and offering drug interaction information.

4. Telemedicine Support:

Virtual assistants can facilitate telemedicine appointments by connecting patients with healthcare providers and assisting with appointment setup and reminders.

5. Healthcare Diagnostics:

AI-powered virtual assistants are being developed to assist with diagnostics by analyzing medical images and patient data.

6. Security and Compliance:

In healthcare, strict adherence to security and privacy regulations (e.g., HIPAA in the U.S.) is crucial to protect patient data when using virtual assistants.

In conclusion, virtual assistants have become integral to various aspects of our lives, from managing our smart homes to improving our productivity at work and enhancing healthcare experiences. While they bring undeniable benefits, they also raise important considerations related to privacy, security, and data ethics. Striking the right balance between convenience and safeguarding personal information is an ongoing challenge as virtual assistants continue to evolve and reshape how we interact with technology.

The Emotional Connection:

The Emotional Connection: Attachment to Virtual Assistants and AI Companions

Artificial Intelligence (AI) has evolved beyond its functional roles and has begun to tap into a deeper aspect of human existence - emotions. The emotional connection between humans and virtual assistants or AI companions is a fascinating and complex phenomenon. This connection not only highlights the advancements in AI technology but also raises intriguing questions about human psychology, ethics, and the future of human-AI interactions. In this exploration, we will delve into the emotional responses people have towards AI, the concept of attachment to virtual assistants, and the role of AI as companions.

1. Emotional Responses to AI:

Human beings are inherently emotional creatures, and our interactions with AI have started to evoke emotional responses. This phenomenon can be attributed to several factors:

- **Anthropomorphism:** People tend to anthropomorphize AI, attributing human-like qualities and emotions to virtual assistants like Siri, Alexa, or chatbots.

- **Convenience and Reliability:** AI's ability to provide immediate responses, assist with tasks, and adapt to user preferences can foster positive emotions like trust and gratitude.

- **Personalization:** Virtual assistants leverage data and machine learning to offer personalized experiences, creating a sense of recognition and connection.

- **Social Interaction:** AI chatbots, especially those integrated into social media and messaging platforms, simulate human conversation, leading to social and emotional engagement.

- **Design and Voice:** The design and voice of virtual assistants can evoke emotional responses. A friendly and empathetic tone can make users feel more connected.

2. Attachment to Virtual Assistants:

Attachment, typically associated with human-to-human relationships, is now extending to virtual assistants. People can form emotional bonds with their AI companions, even though they are aware that these entities are not sentient. Here's why:

- **Consistency and Availability:** Virtual assistants are always available and consistent in their responses, providing a sense of reliability akin to dependable companions.

- **Personal Confidants:** Some users confide in virtual assistants, sharing personal thoughts, feelings, and experiences, which suggests a level of trust and attachment.

- **Dependency:** As AI becomes integrated into daily life, users can become dependent on virtual assistants for tasks and information, leading to a sense of reliance and attachment.

- **Emotional Support:** Virtual assistants can provide emotional support, offering comfort, empathy, and even therapeutic interactions in certain cases.

- **Long-Term Relationships:** Users who have been using virtual assistants for years can develop a sense of attachment due to the long-term relationship they have with these AI entities.

3. Emotional Impact and Ethical Considerations:

The emotional connection between humans and virtual assistants has raised ethical concerns:

- **Manipulation:** There's a fine line between designing AI to elicit positive emotions and manipulating users' emotions for commercial or manipulative purposes.

- **Privacy:** The emotional connection often involves sharing personal information, raising questions about data privacy and consent.

- **Emotional Well-being:** While AI can provide emotional support, it's crucial to ensure that users don't rely on AI for emotional companionship to the detriment of real-life human relationships.

- **Dependence:** Users who become overly dependent on AI companions may face challenges in forming and maintaining meaningful human relationships.

4. AI as Companions:

AI's role as companions goes beyond virtual assistants in smartphones or smart speakers. Companion robots, for instance, are designed explicitly to provide companionship and emotional support:

- **Elderly Care:** Companion robots are used in elderly care facilities to provide companionship and monitor the well-being of residents.

- **Children's Companions:** AI-powered toys and devices are created to be companions for children, fostering social and emotional development.

- **Mental Health:** AI chatbots and virtual therapists offer emotional support and mental health assistance to individuals struggling with various issues.

- **Virtual Pets:** AI-driven virtual pets simulate the companionship and emotional connection that real pets offer without the associated responsibilities.

5. Future Possibilities:

The future holds intriguing possibilities:

- **Emotionally Intelligent AI:** Advancements in emotional AI aim to make virtual assistants more emotionally intelligent, capable of understanding and responding to users' emotional states.

- **Robotic Companions:** Companion robots with advanced AI and human-like features may become more commonplace, offering companionship to a broader demographic.

- **Ethical Guidelines:** The development of ethical guidelines for emotional AI and AI companionship will become increasingly important to ensure responsible use and minimize harm.

- **Human-AI Relationships:** As AI continues to evolve, some individuals may develop deeper emotional connections, sparking discussions about the nature of human-AI relationships and their impact on society.

Conclusion:

The emotional connection between humans and AI is a multi-faceted phenomenon with profound implications for individuals and society as a whole. While AI companions can offer valuable emotional support and assistance, ethical considerations, privacy concerns, and the potential for dependence must be carefully addressed. As AI technology continues to advance, fostering a healthy and responsible relationship between humans and AI companions will be essential to harness the full potential of this evolving field.

Challenges and Controversies Surrounding AI

D ependence, Job Displacement, Data Privacy, Bias, and Fairness
Artificial Intelligence (AI) is a transformative force with the potential to reshape industries and improve our lives. However, its rapid development has also led to various challenges and controversies. In this exploration, we will delve into five prominent issues associated with AI: dependence on AI, job displacement, data privacy and security, and bias and fairness.

1. Dependence on AI:

As AI systems become more integrated into our daily lives, there's a growing concern about over-dependence on these technologies. Several key factors contribute to this challenge:

- **Reliability:** Users may rely on AI systems for critical tasks, assuming they are infallible, which can lead to disruptions when these systems fail or make errors.

- **Skill Erosion:** Excessive reliance on AI for decision-making and problem-solving may erode human skills and critical thinking.

- **Loss of Autonomy:** When AI systems control essential functions, individuals may feel a loss of autonomy and the ability to make independent decisions.

- **Data Dependency:** Many AI systems depend on vast amounts of data, raising questions about data ownership and control.

Balancing the benefits of AI with the risks of dependence is crucial to ensure that these technologies enhance human capabilities without diminishing them.

2. Job Displacement:

The fear of AI causing widespread job displacement is a contentious issue, with proponents and critics alike. Key considerations include:

- **Automation of Tasks:** AI and robotics can automate routine and repetitive tasks across various industries, potentially leading to job loss in those areas.

- **Job Creation:** AI can also create new roles and industries, such as AI developers, data scientists, and robot maintenance technicians.

- **Reskilling:** Encouraging reskilling and upskilling programs is essential to help displaced workers transition to new roles in the evolving job landscape.

The impact of AI on employment is complex, with both challenges and opportunities, making it a topic of ongoing debate.

3. Data Privacy and Security:

The collection and use of personal data by AI systems have raised significant privacy and security concerns:

- **Data Breaches:** AI systems store vast amounts of personal data, making them attractive targets for cyberattacks and data breaches.

- **Surveillance:** AI-enabled surveillance technologies can infringe on privacy rights and be used for mass surveillance.

- **Data Ownership:** Determining who owns and controls the data collected by AI systems is a critical ethical and legal question.

Addressing these concerns requires robust data protection regulations, secure AI systems, and responsible data practices.

4. Bias and Fairness in AI:

Bias in AI algorithms and decision-making processes is a well-documented concern:

- **Training Data Bias:** AI models can inherit biases present in training data, leading to discriminatory outcomes, especially in areas like hiring, lending, and criminal justice.

- **Fairness and Equity:** Ensuring fairness and equity in AI systems is essential to prevent perpetuating societal biases and discrimination.

- **Transparency:** Making AI systems more transparent and accountable can help identify and rectify bias.

Efforts to mitigate bias in AI are ongoing, including the development of fairness-aware algorithms and the establishment of ethical guidelines.

5. Ethical Considerations:

Beyond the specific challenges mentioned above, AI raises broader ethical concerns:

- **Autonomous Weapons:** The development of autonomous weapons systems raises questions about the ethical use of AI in warfare and its potential to escalate conflicts.

- **Emotional Manipulation:** The use of AI in marketing and social media can lead to emotional manipulation and the spread of disinformation.

- **Privacy Invasion:** The invasive capabilities of AI-powered surveillance systems pose ethical dilemmas regarding privacy and civil liberties.

Addressing these ethical concerns requires a holistic approach that involves policymakers, industry leaders, researchers, and the public.

Conclusion:

AI is a powerful tool that has the potential to bring about positive change in society. However, it also comes with significant challenges and controversies that must be addressed. Striking a balance between the benefits and risks of AI involves thoughtful regulation, responsible development practices, transparency, and ethical considerations. The future of AI will depend on our ability to navigate these challenges while harnessing the potential for innovation and progress.

The Future of AI and Human Relationships

Advancements, Scenarios, Ethics, and Preparedness

Artificial Intelligence (AI) is rapidly evolving, transforming how we interact with technology and each other. The future of AI promises to be a dynamic landscape with profound implications for human relationships, society, and ethics. In this exploration, we will delve into the future of AI, advancements on the horizon, potential scenarios, the need for ethical guidelines and regulations, and how we can prepare for the evolving AI-human relationship.

1. AI Advancements on the Horizon:

AI is poised to undergo significant advancements in the near future, driven by ongoing research and technological breakthroughs. Some key areas of advancement include:

- **AI Superintelligence:** Researchers are exploring the development of AI systems with human-level or even superhuman intelligence, which could revolutionize industries like healthcare, finance, and research.

- **Conversational AI:** Natural Language Processing (NLP) and speech recognition technologies are advancing rapidly, making virtual assistants more conversational and capable of understanding context and emotions.

- **AI in Healthcare:** AI will play a pivotal role in healthcare, from diagnostics and drug discovery to personalized medicine and telehealth services.

- **AI in Education:** Personalized learning powered by AI will become more widespread, tailoring educational experiences to individual students' needs.

- **AI in Autonomous Systems:** Advancements in AI will lead to more sophisticated autonomous systems, including self-driving cars and drones, revolutionizing transportation and logistics.

- **AI Creativity:** AI-generated art, music, and literature will become more refined, challenging the boundaries of creativity and artistic expression.

2. Potential Scenarios in the Future of AI and Human Relationships:

The future of AI-human relationships can unfold in various ways, depending on how we navigate the technological landscape. Here are some potential scenarios:

- **Collaborative Partnerships:** AI systems could become collaborative partners, assisting humans in decision-making, creativity, and problem-solving, enhancing productivity and capabilities.

- **Dependency on AI:** There's a possibility of increased dependence on AI for daily tasks, leading to concerns about autonomy and self-reliance.

- **Emotional AI Companions:** AI companions with advanced emotional intelligence may provide emotional support and companionship, especially in situations where human interaction is limited.

- **Ethical Dilemmas:** The rise of AI could lead to ethical dilemmas, such as the use of AI in warfare or the potential for AI to manipulate human emotions.

- **Economic Disparities:** Depending on how AI is implemented, it could exacerbate economic inequalities or contribute to more equitable access to resources and opportunities.

3. Ethical Guidelines and Regulations:

The ethical dimensions of AI are paramount in shaping its future. To ensure responsible AI development, we need clear guidelines and regulations:

- **Transparency:** AI systems should be transparent, and their decision-making processes should be explainable to users.

- **Bias Mitigation:** Robust efforts should be made to mitigate bias and ensure fairness in AI algorithms, especially in contexts like hiring and lending.

- **Privacy:** Stringent data privacy regulations should be in place to protect individuals' personal information and ensure informed consent.

- **Accountability:** Developers and organizations should be accountable for the actions and decisions made by AI systems under their control.

- **Human Rights:** AI should be developed and used in ways that respect fundamental human rights, including freedom of speech and privacy.

- **Global Collaboration:** International cooperation and agreements are crucial to establish a common framework for AI ethics and regulations.

4. Preparing for the Future:

As AI continues to advance, it's essential to prepare for the changing landscape:

- **Education and Training:** Developing AI literacy among the general population and providing training for AI-related professions is crucial.

- **Ethical AI Development:** Developers and organizations should prioritize ethical considerations in AI development and promote responsible AI practices.

- **Public Engagement:** Involving the public in discussions about AI's future, ethics, and regulations ensures that AI development aligns with societal values.

- **Interdisciplinary Collaboration:** Collaboration between technologists, ethicists, policymakers, and social scientists is necessary to address the multifaceted challenges of AI.

- **Continuous Monitoring:** Ongoing monitoring of AI systems and their impact on society helps identify and address emerging issues and risks.

Conclusion:

The future of AI and human relationships holds immense promise, but it also raises complex ethical, social, and technological questions. By advancing AI responsibly, embracing ethical guidelines and regulations, and preparing for the evolving landscape, we can shape a future where AI enhances human capabilities, respects human rights, and fosters collaboration between humans and intelligent machines. The key lies in harnessing the potential of AI while ensuring that its development aligns with the values and interests of humanity.

References

1. **"Artificial Intelligence: A Modern Approach"** by Stuart Russell and Peter Norvig - This renowned textbook offers a comprehensive introduction to AI, covering key concepts, techniques, and applications.

2. **"AI Superpowers: China, Silicon Valley, and the New World Order"** by Kai-Fu Lee - This book provides insights into the global AI landscape, focusing on the competition between China and the United States.

3. **"Weapons of Math Destruction: How Big Data Increases Inequality and Threatens Democracy"** by Cathy O'Neil - It discusses the ethical challenges and potential biases associated with AI and algorithms.

4. **"Rebooting AI: Building Artificial Intelligence We Can Trust"** by Gary Marcus and Ernest Davis - This book explores the challenges of creating AI systems that are trustworthy and unbiased.

5. **"Superintelligence: Paths, Dangers, Strategies"** by Nick

Bostrom - Bostrom delves into the future of AI, discussing the possibilities and potential risks associated with superintelligent AI.

6. **"The Age of Em: Work, Love, and Life when Robots Rule the Earth"** by Robin Hanson - This thought-provoking book explores a future where AI is responsible for emulated human minds.

7. **"AI Now 2021 Report"** - Produced by the AI Now Institute at New York University, this annual report offers in-depth insights into the ethical and social implications of AI.

8. **"Data and Goliath: The Hidden Battles to Collect Your Data and Control Your World"** by Bruce Schneier - While not solely about AI, this book delves into the privacy and security implications of data-driven technologies, including AI.

9. **"The Second Machine Age: Work, Progress, and Prosperity in a Time of Brilliant Technologies"** by Erik Brynjolfsson and Andrew McAfee - It explores the economic and societal impacts of technological advancements, including AI.

10. **"AI in Healthcare: Challenges and Opportunities"** - This research paper from Nature Reviews Drug Discovery discusses the potential of AI in healthcare and the challenges it faces.

Conclusion

A I's Transformative Journey, Key Takeaways, and Looking Ahead

Artificial Intelligence (AI) has embarked on a transformative journey that has reshaped the way we live, work, and interact with the world. From its inception to its current state and the promises of the future, AI's evolution has been both awe-inspiring and challenging. In this comprehensive exploration, we have navigated through the intricacies of AI, delved into its various facets, and contemplated its profound impact on individuals, society, and the future. Let's conclude our journey with a reflection on the key takeaways and a glimpse into what lies ahead in the dynamic world of AI.

Key Takeaways:

1. **AI's Ubiquity:** AI is no longer confined to science fiction; it is an integral part of our daily lives. From virtual assistants and recommendation algorithms to autonomous vehicles and medical diagnostics, AI has become ubiquitous.

2. **Diverse Applications:** AI's versatility knows no bounds. It is applied across diverse domains, including healthcare, finance, education, entertainment, and industry, with each sector harnessing AI's capabilities to solve unique challenges.

3. **Human-Machine Collaboration:** The future of AI is not about humans versus machines but about collaboration. AI augments human capabilities, making us more efficient, creative, and capable in our endeavors.

4. **Ethical Considerations:** As AI continues to evolve, ethical considerations are paramount. We must address issues of bias, fairness, transparency, and privacy to ensure AI benefits society while minimizing harm.

5. **Job Transformation:** AI is changing the landscape of work. While some jobs may be automated, new opportunities are emerging in AI development, data science, and technology-related fields.

6. **AI and Creativity:** AI is pushing the boundaries of creativity, producing art, music, and literature that challenge our perceptions of human versus machine-generated content.

7. **Healthcare Revolution:** AI is revolutionizing healthcare, from early disease detection to personalized treatment plans, offering the potential to save lives and improve patient outcomes.

8. **AI in Education:** In education, AI is reshaping how we learn, offering personalized experiences, adaptive assessments, and innovative teaching tools.

9. **Data's Central Role:** Data is the lifeblood of AI. Quality data is essential for training AI models and ensuring their accuracy and effectiveness.

10. **Human-Centric Approach:** AI's future success hinges on

a human-centric approach. It should be designed to augment human abilities, enhance well-being, and uphold ethical standards.

Looking Ahead:

As we peer into the future, several key trends and developments in AI are poised to shape our world:

1. **AI for Good:** AI will increasingly be harnessed to address pressing global challenges, such as climate change, healthcare disparities, and disaster response.

2. **AI and Climate Change:** AI's data analytics and modeling capabilities will play a pivotal role in mitigating and adapting to the effects of climate change.

3. **AI Regulation:** Governments and international bodies will continue to develop and enforce regulations to ensure the ethical and responsible use of AI.

4. **AI Ethics and Accountability:** As AI becomes more sophisticated, discussions surrounding AI ethics and accountability will gain prominence, guiding its development and use.

5. **AI in Space Exploration:** AI will facilitate space exploration by aiding in autonomous spacecraft navigation, data analysis, and even supporting astronaut health.

6. **Quantum AI:** The convergence of quantum computing and AI will open up new horizons in computing power, potentially accelerating AI research and applications.

7. **AI-Powered Healthcare:** AI-driven healthcare will become

increasingly personalized, with AI assisting in early diagnosis, treatment recommendations, and drug discovery.

8. **AI in Education 2.0:** AI will continue to transform education, offering advanced tools for remote and personalized learning, bridging educational gaps worldwide.

9. **AI and Augmented Reality (AR):** AI-driven AR experiences will become more immersive and integrated into our daily lives, enhancing how we perceive and interact with the world.

10. **AI in Governance:** Governments will leverage AI for improved governance, data-driven decision-making, and better public services.

In conclusion, AI's journey has been a testament to human ingenuity and innovation. It has brought both promise and challenge, but its potential to enhance our lives and tackle global issues is undeniable. As we navigate the evolving landscape of AI, it is imperative that we maintain a commitment to ethical principles, collaboration, and responsible development. The future of AI is a shared endeavor, and together, we can shape a world where AI serves as a force for good, enriching our lives and advancing humanity.

www.ingramcontent.com/pod-product-compliance
Lightning Source LLC
Chambersburg PA
CBHW062314290526
45794CB00005B/1805